HOW TO MAKE YOUR OWN
ALCOHOL
FUELS
BY LARRY W. CARLEY

TAB TAB BOOKS Inc.
BLUE RIDGE SUMMIT, PA. 17214

FIRST EDITION

FIRST PRINTING—AUGUST 1980
SECOND PRINTING—NOVEMBER 1980
THIRD PRINTING—JANUARY 1981
FOURTH PRINTING—MARCH 1981
FIFTH PRINTING—MAY 1981

Library of Congress Cataloging in Publication Data

Carley, Larry W
    How to make your own alcohol fuels.

    Includes index.
    1. Alcohol as fuel. I. Title
TP358.C38   662'669   80-14459
ISBN 0-8306-9931-7
ISBN 0-8306-2074-5 (pbk.)

Foreword

The purpose of this book is to provide the reader with the basic background information necessary for the home production and use of alcohol fuel. Using this information, the do-it-yourselfer can design and build a still and make the appropriate modifications to the equipment so that it will run on alcohol fuel. For this reason, the author and publisher assume no responsibility or liability for the use or misuse of such home-made distilling equipment, alcohol fuel or modifications to vehicles or equipment.

It should also be noted that this is not a moonshiner's manual. The procedures and equipment described in this book are for the home production of fuel alcohol only—not drinking alcohol. Fuel grade alcohol might contain various contaminants that can be harmful or fatal if taken internally. It is against the law to manufacture drinking alcohol without a special permit from the Bureau of Alcohol, Tobacco & Firearms. Failure to obey the law can result in severe criminal penalties.

Larry W. Carley

Contents

Introduction

This is a book about alcohol fuel. It's also a book about self-sufficiency, economic opportunity and good 'ol American ingenuity.

Each passing day brings with it a greater concern over the future cost and availability of gasoline and diesel fuel. Nobody can predict the eventual outcome of our present energy crisis or how much we will be paying for fuel next year or the year after—or even if we'll have fuel available at any price. This is why we need to pursue energy alternatives now.

After talking with countless individuals who are involved directly and indirectly with the alcohol fuel movement, alcohol production, still manufacturing, engine conversions, energy research, fuels research and agricultural equipment manufacturing, I'm convinced alcohol is the best alternative we have.

Alcohol is the one fuel that comes closest to being the ideal replacement for gasoline.

Not everybody agrees that alcohol is the way to go. A wealth of opinion about the value of alcohol as a fuel is sharply divided. Some people are stubbornly against it without good reason, for selfish reasons, because they're misinformed or because they believe the anti-alcohol propaganda circulated by certain special interest groups. Other people are staunchly in favor of using alcohol fuel for no good reason, for selfish reasons, because they believe the pro-alcohol propaganda—or because they've examined the facts and reached an intelligent conclusion.

The controversy over alcohol fuels stems from our economic dependence on a single fuel resource—petroleum. And any threat to the established order is bound to meet with stiff resistance.

Considering alcohol as a viable alternative to petroleum means changing the way we think about energy production. It means changing from a non-renewable fossil fuel to a resource we can grow for ourselves. It means eliminating our dependence on others to fulfill our basic energy needs.

If the people of this country agree on one thing about the energy crisis, it's that we have a serious problem. Whether there is a real energy shortage or a contrived shortage is not the point. We have a crisis because we've put all our eggs in one basket. Oil is a limited resource and a resource that can be manipulated by a handful of men—be they the oil ministers of OPEC, the executives who sit in the boardrooms of Big Oil or the politicians and bureaucrats who decide the policies and regulations that assure our continued dependence on oil.

We spend all our time arguing about who to blame for the energy crisis when the real culprit is our total dependence on a single energy resource.

We've developed such a gluttonous appetite for oil because it was traditionally the cheapest fuel to burn. It cost less than other alternatives so we built our economy upon it. Once we were totally dependent, guess what happened? The price shot straight up—a textbook example of supply and demand. OPEC and Big Oil have the oil and we need it so we'll pay anything to get it. If you don't agree, stop and think about how much you're paying for fuel now compared with just a few years ago. Convinced?

It's frightening when you stop to think about our dependence on this single resource. American mechanized agriculture must have gasoline and diesel fuel to plant, cultivate, harvest and transport crops and livestock. In addition, there are all the fertilizer, pesticides and other agri-chemicals that are derived from petroleum and natural gas. All truck and rail transportation in this country, which is the lifeblood of our economy, depends upon an uninterrupted supply of diesel fuel. All air travel for business and pleasure is powered by aviation gas or kerosene (from petroleum). All inland waterway shipping, vital to the transport of grain and bulk commodities (including coal for power plants) depends on diesel power. Even our export of agricultural and industrial products to the rest of the world depends on diesel powered shipping.

Needless to say, driving 55 mph, car pooling, making fewer trips to the store and cutting back on pleasure driving is a drop in the barrel compared with our total consumption of petroleum derived fuels. Conservation does nothing to change our dependence on oil. It only prolongs the agony.

What we need are cheap, safe, diversified energy alternatives.

What alternatives do we have? Ask a nuclear engineer and he or she will tell you nuclear is the only way to go. Obviously, you can't power a tractor, car or truck with a nuclear reactor but you can build nuclear generating plants to make electricity. The electricity can then be used for industry and for charging battery-powered vehicles. The idea is to reduce some of our dependence on oil. But stop and look at the consequences. Besides a highly questionable track record of operating safety and the potential for a serious environmental disaster, nuclear energy requires the indefinite storage of highly radioactive wastes that will remain deadly for thousands of years. Nobody wants those wastes buried in their backyard. Future generations will have enough problems without inheriting a nuclear timebomb.

But what's worse is that nuclear energy is just trading one kind of dependence for another. Like petroleum, it's high technology. It's beyond the average person's ability to comprehend. It's a multi-billion dollar corporate scale venture subject to the same kind of manipulation and control prevalent in the petroleum industry.

Building dozens of additional nuclear generating plants like those at Three Mile Island is not an acceptable or realistic answer. Who could guarantee an uninterrupted flow of uranimum ore from abroad to fuel these additional plants? Would the uranium producing nations form a cartel once our total dependence on nuclear energy was assured? You bet they would.

Breeder type reactors are not the answer either. These are a class of reactor that "breeds" or makes more fuel than they use. Sounds like something for nothing, doesn't it? The catch is that breeder reactors still have all the same operating hazards as regular reactors, plus they make large amounts of plutonium isotopes. Plutonium, in case you didn't know, is one of the deadliest and most toxic radioactive wastes known. It remains radioactive for millions of years, it is harmful to life in concentrations as small as a few parts per billion, and it is also the preferred ingredient for making nuclear weapons. So much for breeder reactors.

Another so-called solution to the energy crisis is synthetic fuel. Big Oil, coal interests and others propose to solve our long-term energy needs by keeping us dependent on gasoline and diesel fuel. They'll make the stuff synthetically from coal, natural gas, oil shale, tar sands or heavy crudes using the wizardry of modern chemistry. Whether or not synthetic fuels will prove to be economical is not as important a question as who will control the resource. High technology synthetic fuel production is still a monopoly enterprise—beyond the control of those who would be dependent upon it. There are also major environmental questions to be resolved. Should we strip mine large areas of the western United States and poison their limited water resources for the sake of a few more gallons of gas, or should we rethink our dependence on oil and seek other alternatives? I opt for the later.

There are a number of diversified energy sources easily within our grasp. These include alcohol, solar energy, wind power, geothermal, methane and hydrogen. Each has certain advantages and disadvantages. The method I believe to be the most practical and timely in terms of reducing our dependence on petroleum is alcohol.

Solar energy is fine for heating buildings, making hot water and generating electricity. Except for very specialized applications, electrically powered vehicles and farm tractors are as yet extremely impractical. The biggest problem with electric vehicles is storing sufficient energy in a small and lightweight space. This is something batteries do poorly but liquid fuels do quite well. Liquid fuels can be thought of as liquid energy. Unfortunately, you can't think of batteries as being solid energy. They're mostly bulk. They're also heavy and costly. Even some of the more exotic designs on the drawing board are still a long way from the energy density of gasoline or diesel fuel.

The typical electric vehicle of today is heavy, sluggish on acceleration, and has a very limited range. General Motors has taken a Chevette, crammed it full with about 1000 pounds of batteries, and called it the car of the future. It's lucky to make 100 miles on a charge.

The same car with a conventional gasoline engine can cover 100 miles on less than 4 gallons of gasoline—and keep right on going.

Another problem is that batteries must be replaced every few years or about every 20,000 to 30,000 miles. Replacement isn't exactly cheap. Batteries also suffer from corrosion problems and this makes preventative maintenance something of a chore. Batteries also take hours to recharge whereas a fuel tank can be refilled in a matter of minutes. Do you still think electrics are the coming thing? Not for a long time to come. Liquid fuels are much more efficient and convenient.

Wind power is another alternative often discussed. Like solar, it is ideal for generating electricity. Unfortunately it is useless for powering vehicles with conventional gasoline or diesel engines. For the wind to be of any use, a vehicle must first be converted to electric power at considerable cost. The wind can then be used to turn a generator to charge the batteries. But why go to all that effort when you can make a replacement fuel that will burn in your existing engine?

Geothermal energy is another possibility. It too, can be used to heat buildings or generate electricity. But unless you've got a hot spring or volcano nearby, forget it.

What about hydrogen, the "clean burning fuel of the future?" Hydrogen can be extracted from water by splitting water molecules into hydrogen and oxygen with electricity. This process is called electrolysis. It's not terribly difficult to do but most hydrogen projects have failed to show favorable economics. The process is still too costly to reasonably compete with other alternatives. It makes more sense to use the electricity directly than to use it to make hydrogen.

Hydrogen is also a highly explosive gas. Remember the Hindenburg? It was full of hydrogen as it crashed. Handling hydrogen requires special precautions as well as the use of pressurized storage tanks. A gasoline engine can be converted to run on hydrogen with good results, but not as easily as it can be converted to run on alcohol.

How about methane? Like alcohol, methane is a home-grown fuel that can be made from biomass or other organic wastes such as manure, grass clippings or even municipal garbage. Methane is fairly simple to make and it is an ideal fuel for heating, cooking or powering stationary engines. For use in a vehicle or farm tractor, it requires the addition of a pressurized fuel system and a modified carburetor capable of handling the gaseous fuel.

Are there any other alternatives? Some people have suggested a return to animal power or human muscle power. True, we can reduce our dependence on machines somewhat—but try farming 900 acres with a mule and plow. Such a philosophy might work in developing nations where agriculture and transportation have not yet become as energy intensive as our own. Returning to such a way of life would be a monumental undertaking that would require drastic changes in lifestyles, technology and society itself. It's doubtful the population of an overcrowded world already plagued with malnutrition could survive without mechanized agriculture.

The best alternative to megabuck technology—spending more to receive less, overdependence on a limited resource or a potential collapse of our entire way of life—is to develop the abundant energy resources we already have. All the cheap oil is gone. The only way to sustain our petroleum diet is to drill deeper, search out every marginal oil field and spend millions figuring out new ways to make oil from something else. Petroleum has lubed the wheels of our economy quite well but the honeymoon is over.

Alcohol alone won't solve the energy crisis, but it is certainly a step in the right direction. The more alternatives that we develop, the more balanced our energy diet will become.

Alcohol is a fuel you can make yourself. It's based on techniques thousands of years old and on equipment you can build yourself. It's a natural substitute for petroleum fuels in many instances and it is a viable alternative to the vicious cycle of higher prices and uncertain supplies.

Every gallon of foreign oil we continue to buy underlines our dependence on a non-renewable resource and further weakens our position in an energy-starved world.

Every gallon of oil bought at inflated prices, whether from foreign sources or our own domestic producers, further fuels the fires of inflation.

But every gallon of alcohol fuel we make for ourselves is one gallon closer to self-sufficiency and a new way of life.

The choice is yours.

Chapter 1
The Truth About Alcohol

The argument heard most often against using ethanol alcohol as a motor fuel, either straight or blended with other fuels, is that it just isn't practical. Opponents say alcohol won't work for a number of reasons: It's too difficult and time-consuming to make. We need our surplus crops for food, not fuel. Alcohol costs too much and will never be able to compete economically with gasoline, diesel oil or other synthetic fuels. Alcohol makes a poor engine fuel and causes more problems than it cures.

It seems that the anti-alcohol forces have no shortage of energy when it comes to criticizing what might be our only viable alternative to OPEC, Big Oil, rising prices and the increasing scarcity of conventional petroleum fuels. Whether or not alcohol will prove to be America's answer to OPEC depends on too many variables. I'm not about to tackle all the issues surrounding the future of alcohol, the feasibility of a national gasohol program and so on because these subjects are beyond the scope of this book. But one thing is for sure: Alcohol is a better answer to the energy crisis than the lack of answers we've had so far. With that in mind, consider some of the criticisms of alcohol and see how accurate they really are.

CRITICISM: MAKING ALCOHOL
IS TOO DIFFICULT AND TIME-CONSUMING

If you think making alcohol sounds tough, try making your own gasoline. As you'll discover in the following chapters, making alcohol is a proven and relatively straightforward process that is well within the abilities of most people. The key to making alcohol is a sound understanding of the procedures involved and the right equipment to handle the job. Anybody can build a still and anybody can mix the necessary ingredients (Fig. 1-1). Knowledge and skill are what separate the successful from the not-so-successful when it comes to making alcohol.

Like baking a cake, you don't just stir up the recipe, throw it in the oven and forget it. You have to control things such as time and temperature if you want the cake to turn out right. The same applies to brewing alcohol. That means controlling things like sugar content, pH and temperature. These are some of the aspects of the science of making alcohol that must be learned. I hope this book will guide you through the basics and help you avoid the pitfalls most beginners are likely to encounter.

The most time consuming part of making alcohol is the fermentation process. This is when the yeast is added to the mash to transform the sugars into alcohol. Fermentation takes several days to complete—but the yeast do all the work. Your only concern during this period is for the temperature within the fermentation vat. It shouldn't be allowed to rise too high, otherwise it will stop the yeast action. The solution is to install a simple automatic thermostat control that will regulate vat cooling.

You can spend as much time making alcohol as you wish, but the addition of a few simple controls can cut the time spent "babysitting" a still to an absolute minimum.

The single most important factor in efficient fuel production is building a still with sufficient capacity. The larger the still, the fewer the number of runs that are necessary to keep your fuel tanks filled. If a still is the proper size, you need only make alcohol an average of once a month. That's a schedule anyone should be able to live with.

CRITICISM: WE NEED OUR SURPLUS CROPS FOR FOOD

What do we have a shortage of in this country, food or fuel? If you guessed food, guess again. The United States usually produces about 2.5 billion bushels of surplus grain annually and takes another 1.3 billion bushels of grain out of production each year as part of the U.S. Department of Agriculture's set-aside program. We have a tremendous agricultural potential in this country. As a result, our over-production forces the government

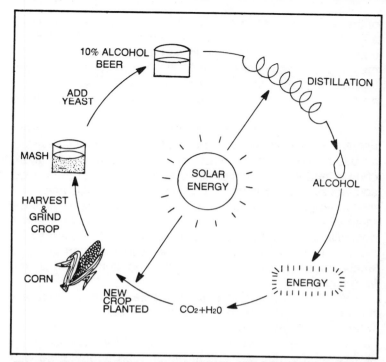

Fig. 1-1. Energy cycle for a renewable energy resource such as corn. Through the process of photosynthesis, plants take atmospheric CO_2 and water to build sugars, starches and cellulose. The crop is then harvested, ground and made into mash. Yeast is added to the mash to ferment the sugars into alcohol. Solar energy can then be used to distill the alcohol from the beer. When the alcohol is burned as fuel, the stored energy is released and the waste products are returned to be used over again.

to support crop prices. Year after year, farmers are faced with fluctuating markets and prices that always seem to be far less than they should be. And year after year, more farmers are calling it quits. Using only a portion of our great surplus could bring change to that.

Even the Department of Agriculture has indicated "There is a good argument for supporting grain prices by maximizing production and converting surpluses to ethanol fuel, rather than by idling land."

What about the food shortage in the rest of the world? There's no shortage of hungry mouths to feed and little shortage of carbohydrate to fill bloated stomachs. The real shortage is protein. Making alcohol only involves the carbohydrate portion of the grain and not the protein. Only the sugars and starches are fermented into alcohol. This leaves a high-protein concentrate called distillers dried grain and solubles (DDGS). Commercial distilleries package and sell this product as an animal feed supplement or for human consumption. This same residue from your still can be fed wet to livestock or spread on fields as fertilizer.

Another fact to consider is that alcohol can be made from damaged crops and waste materials that might otherwise never be used. These include citrus fruit peelings, mildewed grain, cheese and whey. Alcohol can also be made from non-food sources such as corn stalks, plant leaves, wood pulp and even municipal garbage. Although converting such high cellulose materials into alcohol really isn't practical on a small scale, it does illustrate the fact that there are abundant resources for making alcohol all around us. These resources can be used without producing a negative impact on our human and animal food supply.

What's more, alcohol is a renewable resource. There are no wells to run dry and no mines to peter out. We can grow what we need year after year.

When gasoline or diesel oil, is burned we are using up stored energy that was created long, long ago. Not so with alcohol. The living plants from which alcohol is made represent new energy. They are the on-going product of the sun, the earth and photosynthesis.

Plants use sunlight to convert moisture, atmospheric CO_2 and nutrients in the soil, into starches, sugars and cellulose. The end products of photosynthesis can be harvested and made into alcohol.

There is a real concern among scientists that we might seriously upset the earth's balance of atmospheric CO_2. Burning fossil fuels generates great quantities of CO_2. One theory suggests that at our current rate of energy consumption, the accumulation of excess CO_2 will eventually create a "greenhouse effect." The earth will retain more heat than it normally radiates into space. The result will be a gradual rise in worldwide temperatures. This could melt the polar icecaps, upset global weather patterns and have a disasterous impact on world crop production.

Obviously, switching more of our energy emphasis over to alcohol won't solve the problem, but it will give us more time to make the transition from the fossil fuel age to the coming solar age.

CRITICISM: ALCOHOL TAKES MORE
ENERGY TO PRODUCE THAN IT YIELDS AS A FUEL

If you stop to add in all the costs of finding, drilling, transporting, refining and marketing that go into every gallon of gasoline and diesel oil, then these fuels also show a net energy deficit.

Most of the comments regarding the so-called "poor" efficiencies of alcohol production are based on the energy balances of existing distilleries. However, these plants were built to brew expensive liquors, not cheap motor fuel. Drinking alcohol requires very careful preparation as well as additional distilling to enhance the flavor, appearance and physical properties of the bottled product. These steps are totally unnecessary for a motor fuel. Most distilleries were built years ago when energy costs and recycling were not as much of a concern as they are today. Newer, more sophisticated distilling equipment with heat recovery systems have been proved highly energy efficient in large-scale plants.

But what about home-built stills? Even a relatively crude design can be energy efficient if the heat source for cooking and distilling comes from solar energy or waste materials. Burning corn cobs, husks, stalks or other refuse increases the net energy balance of the process by using materials that would otherwise go to waste. Such materials are a lot cheaper to burn than oil, natural gas or propane.

On-farm stills have another advantage over the large distilleries. They are closer to the source of raw materials. Fully 60 percent of the operating cost of a commercial alcohol plant is the purchase and transportation of raw materials. The farmer doesn't have this problem. His only cost is the energy expended to plant, cultivate and harvest the crop.

A small scale alcohol production facility can also be built close to a source of recyclable waste material such as whey from a cheese plant, fruit peelings from a fruit packing plant or spoiled grain from a grainery. The raw materials for alcohol production can be the left-over wastes of other industries. Under such circumstances, alcohol production can become a byproduct of the primary industry. Overall operating efficiency improves because materials that would otherwise be discarded as waste are recycled into a valuable energy resource. The alcohol can be sold for a profit or used to power plant equipment and vehicles for additional savings.

CRITICISM: ALCOHOL CANNOT
COMPETE ECONOMICALLY WITH GASOLINE OR OTHER FUELS

This argument is based on the assumption that the raw materials that go into making alcohol will always be higher priced than the raw materials from which gasoline, diesel fuel and synthetic fuels are made.

The roots of this question go back to the turn of the century. When the first Model A's came out, they were equipped with adjustable or dual feed carburetors so that farmers could use either gasoline or alcohol, depending on which fuel happened to be available in their area. At this point in history, alcohol and gasoline were opposing contenders for the lucrative motor fuel market that was about to blossom. The age of the automobile was just beginning and it was obvious that the winner would reap enormous profits from the emerging industry.

As it turned out, a number of factors quickly destroyed the chance alcohol had for becoming the national motor fuel. For one thing, alcohol had (until 1906) the added burden of a 40 cents per gallon "liquor" tax. This made alcohol far more costly than gasoline. Alcohol's biggest disadvantage, however, was the relative weakness and disorganization of the industry. It consisted primarily of a few scattered family-owned distilleries. Gasoline, on the other hand, enjoyed the combined forces and fortunes of the Standard Oil Trust and John D. Rockefeller. Needless to say, the fledgling alcohol industry never had a fighting chance. By 1911, gasoline easily monopolized the motor fuel market. About the same time, the Western oil boom began, making abundant supplies of cheap petroleum readily available to the rapidly expanding gasoline industry. The age of the internal combustion engine and Big Oil was off and running.

In the 1930's, during the depths of the Great Depression, there was a brief resurgance of interest in alcohol fuel. Farm prices hit rock bottom and new markets were desperately needed to help boost the depressed markets. Supporters claimed alcohol was the answer to the economic woes of the nation. There was a lot of talk, a lot of debate and a lot of anti-alcohol propaganda spread around by Big Oil. After a brief flurry of publicity, vested interests won out and the movement fizzled.

Alcohol received its biggest boost during World War II when it became a vital industry for the war effort. When our supplies of natural rubber were cut off by the Japanese, we needed alcohol to make synthetic rubber. As the war continued, production increased dramatically and alcohol found its way into gunpowder, medicines, aviation fuel and "torpedo juice." When the war ended, the government dropped its support of the wartime alcohol industry and the distilleries closed. Cheap petroleum was once again available and Big Oil profits were better than ever.

World politics took another turn in the early 1960's. Within a few years, a new wave of nationality swept across Africa and the Middle East. The Middle Eastern oil fields that had been drilled and developed by Big Oil were gradually coming under the control of the Arabs. This was considered a minor concession to Arab sovereignty and few people worried very much about the long term consequences. When the 1973 Arab oil embargo came along, Americans suffered a rather rude awakening. OPEC was front page news and the era of cheap energy was suddenly over. Over the years, the importing nations of the world have become more and more dependent on oil from the Middle East.

As it stands now, new alcohol plants reportedly coming on stream will be able to produce fuel grade alcohol for 60 cents to $1.50 per gallon, depending on the source and prevailing price of the raw materials. These large scale commercial ventures will be competing head on against gasoline and diesel fuel. They will mean a new market for "energy" crops and an increased supply of alcohol for gasohol or alcohol powered vehicles.

As OPEC price hikes continue, the economics will become increasingly favorable towards alcohol fuels. Of course, this will also aid the economics of making synthetic gasoline from coal. However, that is something you can not do yourself.

What about the economics of a backyard still? As the technology stands today, do-it-yourself alcohol can compete with existing prices for petroleum fuels. How well alcohol fuel competes depends on the efficiency of the still, the heat source that powers the still, the raw materials being used and the current market prices for those raw materials.

Something very important to remember when figuring a cost comparison is that the bulk of a grain crop remains as a high protein food supplement after the alcohol has been extracted. If the feed value of this byproduct is figured into the cost equation, the economics become quite attractive. In fact, the large commercial distilleries package and sell this DDG (distillers dried grain) as a high-protein animal feed supplement. University research studies have proven that animals actually gain weight faster on a ration of DDG and grain than on grain alone.

CRITICISM: ALCOHOL MAKES A POOR ENGINE FUEL

This myth is based on alcohol's "reputed" poor performance when it comes to mileage, fuel system corrosion and cold weather starting.

True, pure alcohol generally does not deliver as good fuel mileage as gasoline because ethanol contains fewer Btu's per gallon. Gasoline averages somewhere around 120,000 Btu's per gallon whereas ethanol has only about 88,000 Btu's. The actual difference in observed mileage, however, when an engine is converted to alcohol varies from *zero* to as much as 50 percent worse. It all depends on the application and the type of modifications made to the engine and fuel delivery system.

For example, two engineers from Elgin, Illinois actually achieved in improvement in fuel mileage when they converted a 1975 Pinto to run on 180 proof ethanol. They did it by increasing the compression ratio of the engine and substituting a vapor injection system for the standard carburetor.

Ethanol's higher octane rating, about 10 to 14 points higher than an average unleaded gasoline, means the compression ratio of the engine can be raised to as high as 12:1 for greater combustion efficiency. This translates into more power and better mileage.

Since you can't run straight alcohol in a gasoline engine without compensating for the difference in required air/fuel ratios (alcohol needs a richer mixture), an unmodified engine will naturally deliver less power on alcohol. Open up the carburetor jets, advance the timing to compensate for the higher octane of ethanol and the power comes right back.

The same holds true for blends of alcohol and gasoline. When the proportion of alcohol in the blend exceeds about 20 percent a "leaning effect" in the air/fuel ratio results in a noticeable loss of horsepower. Does this mean "gasohol" (a blend of 10 percent ethanol and 90 percent unleaded gasoline) delivers fewer miles per gallon than ordinary gasoline? Not at all.

According to the results of a two-million mile road test conducted by the University of Nebraska, gasohol showed an average overall increase in mileage of 5 percent compared to gasoline. The tests were conducted on state owned vehicles under everyday driving conditions.

Mileage and performance differences with gasohol depend entirely on the application. Because of the 10 percent content, gasohol creates a leaning effect in the air/fuel ratio of an unmodified carburetor. This leaning effect is so slight as to be unnoticed in normal driving, but it does have an effect on overall mileage.

On cars built in the 1960's, with rich air/fuel ratios, the leaning effect produced by gasohol generally produces an improvement in fuel mileage. On cars made in 1973 and 1974, with excessively lean carburetor calibration, to meet the tough Federal anti-pollution laws, a slight loss of mileage might result. Cars built since 1975 might or might not pick up some extra mileage on gasohol—depending on how the engine is calibrated at the factory.

On many of the newer cars with electronic feedback systems, such as GM's C-4 closed loop or Ford's EEC-11 and EEC-111, an oxygen sensor in the exhaust manifold compensates for any leaning effect by signalling the carburetor to maintain the correct air/fuel ratio. This means that many of the cars that will be built in the 1980's might be able to run on much higher mixtures of ethanol than the current 10 percent now found in gasohol.

Another factor that might increase gasohol and alcohol's popularity with the Federal government is that ethanol is a clean burning fuel. Carbon monoxide and hydrocarbon emissions are significantly lower than those for gasoline. Evaporative emissions are somewhat higher, but are not considered as much of a serious threat to the environment as the exhaust that comes out the tailpipe.

What about other so-called problems? Alcohol has been blamed for fuel system corrosion because it absorbs moisture from the air and it is an organic solvent.

As for absorbing moisture from the air, this is not a problem in modern sealed fuel tanks. Because water does mix quite readily with alcohol, you don't have to worry about wintertime gasline freeze. In fact, alcohol is what is used in most of the commercially sold "gasoline antifreezes" you dump in your fuel tank.

Unfortunately, alcohol is a rather strong organic solvent. This means it might soften some types of plastic found in fuel filters, fuel lines, fuel pumps or carburetor floats or gaskets. Alcohol also tends to dissolve the varnish that accumulates in gasoline fuel tanks and this might in turn clog the fuel filter the first time alcohol or gasohol is used. Fortunately, such problems have been rare with the 10 percent blend of gasohol currently being sold. In fact, the car manufacturers have enough confidence in gasohol that they have extended their new car warranties to cover use of the fuel.

As for straight alcohol conversions, problems with most quality plastic and rubber materials are unlikely. If they can withstand gasoline, they can usually withstand alcohol. However, there are exceptions. If a problem is encountered, the cure is easy enough. Simple replace the faulty component with one made from a material that is compatible with alcohol. For example, you could replace a plastic carburetor float with a brass one.

As for the question of cold weather starting on straight alcohol fuel, any problems can be overcome a number of ways.

● Start the engine on gasoline and then switch over to alcohol once the engine warms up.
● Use a booster squirt of gasoline or ether to assist cold morning starts.
● Install a simple electric fuel preheater to warm the fuel so it will vaporize more readily. If you live in a warm climate you won't have this problem.

Alcohol has different physical properties, such as being safer to handle and store because it is less explosive, and different performance characteristics, cleaner burning, higher octane, than gasoline. Its use, therefore, is well within the technical abilities of most do-it-yourselfers and involves only those modifications which are necessary to take advantage of its unique properties. What about diesohol? Blends of up to 50 percent alcohol have been run, but usually with disappointing results. Because alcohol contains less heat value than diesel fuel, a loss of horsepower is usually experienced. Alcohol is also a much thinner fuel than diesel oil, meaning lubrication problems can develop within a standard diesel injection pump. Unless the pump is specifically designed to handle the lighter fluid, damage might result.

Another problem with burning a blend of alcohol and diesel oil, or straight alcohol in a diesel engine, is that ethanol has a low cetane number, exactly the opposite of what is needed for a compression ignition engine. Because of these differences, many "experts" believe alcohol is a better fuel for spark ignition engines than diesels.

Alcohol has been used successfully in diesel engines when mixed with vegetable oils or when fumigated into the intake manifold as a supplemental fuel. When used as a supplemental fuel, significant improvements in performance and horsepower have been shown, along with a big savings in diesel fuel consumption.

Converting a diesel engine is not a simple process, but it can be done. A diesel engine can be engineered from the ground up to run on alcohol, a blend of alcohol and diesel or on a dual fuel.

If you're still not convinced alcohol is a worthy fuel, consider the Indianapolis 500. The 33 cars that compete in the world's fastest auto race all burn alcohol (Fig. 1-2). The exact fuel is methanol, ethanol's first cousin. They use alcohol instead of gasoline in their high compression, turbocharged engines because they can get more power out of the alcohol. The Indy 500 race car is a special machine that operates under the most severe circumstances you can imagine. You can bet the mechanics and drivers aren't going to take any chances on running a second rate fuel.

ALCOHOL FUEL AND OUR GOVERNMENT

President Carter asked Congress to approve $11 million in grants, low interest loans and loan guarantees for the construction of 100 small-scale alcohol plants. Congress replied by

Fig. 1-2. Indy 500 race cars burn alcohol fuel because more total horsepower can be achieved on alcohol than with gasoline.

increasing the amount to $500 million. Although this is still only a fraction of the $5 billion Brazil has committed to its alcohol fuel program, it does show our leadership is headed in the right direction.

President Carter asked Congress to extend the 4 cents a gallon excise tax exemption on gasohol past its October, 1984 expiration date. Many states have already reduced or eliminated their excise taxes on gasohol to help it compete more favorable with gasoline pump prices.

Congress has directed the Pentagon to buy as much domestically produced alcohol and gasohol as it can. A similar measure has been ordered for the U.S. Postal Service. Together, these government agencies account for 80 percent of our government's gasoline purchases, or about 375 million barrels of gas a year.

According to a report issued by the Department of Energy on alcohol fuels:

"The Department of Energy is committed to helping alcohol fuels achieve their potential in the Nation's energy future. Alcohol fuels can contribute to U.S. energy resources by using domestic renewable resources to extend supplies of existing fuel. The Department also encourages the production and use of alcohol fuels and urges the States and private sector to continue efforts to develop and market alcohol fuels . . ."

" . . . The Department of Energy will also continue to take steps to eliminate institutional barriers that impede alcohol fuels development . . ."

If that isn't encouraging, consider the role of the Department of Energy: To assist industrial development by supporting existing investment incentives and other techniques to help the private sector develop rapidly; and to provide research and development to help build the foundation on which alcohol fuels can compete effectively with other fuels in the 1980's and beyond.

The Department of Agriculture has indicated that constructive action will be taken to insure land and feedstocks are available to support fuel alcohol production. The USDA has already approved a $30 million loan guarantee program for two new alcohol plants. One is designed to produce ethanol from sugar cane using a new conversion technology. The other is intended to demonstrate the commercial use of sweet sorghum as a feedstock for ethanol production.

The USDA is also funding research into testing new crop varieties for use as ethanol raw materials and research into the design of small scale on-farm technologies for alcohol production.

The Community Services Administration is funding small-scale community alcohol energy projects and the Economic Development Administration will be administrating the bulk of the alcohol funds for the construction of small-scale plants.

All in all, the alcohol fuel movement has achieved the momentum necessary to change the energy outlook in this country—thanks to Big Oil, OPEC, greed, inflation, government inaction, government action, new technology, Mother Nature and our addiction to petroleum.

The truth about alcohol is that it is practical. It offers self-sufficiency and great rewards to those who seize the opportunity to get in on the ground floor of this emerging technology. Farms might soon be thought of as "energy acres" and whole new industries might evolve around energy crops, conversion processes, still designs and equipment, engine conversions and other uses of alcohol.

Chapter 2
Brewer's Basics

Making alcohol is a process that dates back about as far as civilization itself. The Romans knew how to do it, as did the Egyptians, the Phoenicians and even the barbarians. Archaeologists have unearthed simple clay stills from ancient Mesopotamia, the "cradle of civilization."

Of course, they didn't use alcohol in 3500 B.C. for chariot fuel. More likely, it was used by the charioteer the night before the big battle or to celebrate his return should he be lucky enough to survive. Civilization didn't need alcohol for fuel then because horsepower was still the end product of the amount of oats consumed by a horse. It wasn't until the invention of the internal combustion engine that anyone gave alcohol a second thought as an energy source. Until that time, burning it would probably have been considered sacrilegious.

Alcohol's long history as an intoxicating beverage has given mankind many years to perfect the process of brewing. Although there are significant differences between brewing liquor and brewing fuel alcohol, both share the same basic technology. Yeast is added to a simple sugar solution where it is allowed to ferment the sugar into ethanol.

With fuel production, the process is much more basic since flavor, color and cleanliness of the finished product are not important. Because of this, practically any raw organic material can be used to make fuel alcohol. The list of materials includes corn, grain, sugar cane, sugar beets, potatoes, sorgum, and even such high-cellulose materials as plant stalks, leaves, hay, wood pulp or municipal garbage (which is about 35 percent to 55 percent paper). For the do-it-yourselfer, however, high-cellulose materials are not as practical as the feedstocks which contain sugar and starch.

ALCOHOL

Before proceeding to the actual steps of making alcohol, there are a few things you should know about alcohol. Although the word "alcohol" is used throughout this book, the specific type of alcohol is actually ethanol alcohol.

Alcohol actually refers to a whole group of similar chemicals, such as ethanol, methanol and isopropyl. The only two that are presently economical to consider for use as motor fuel are ethanol and methanol. Ethanol is the type of alcohol that can be made by the do-it-yourselfer through the fermentation of sugars from a basic feedstock such as corn or grain. It's the same type of alcohol that is found in liquors, and can also be called grain alcohol, ethyl alcohol or spirits. Methanol, on the other hand, is highly poisonous and is currently produced on a commercial basis from coal, natural gas or petroleum. Alcohols share the general formula of:

$$C_n H_{2n+1} OH$$

where C stands for carbon atoms, H for hydrogen atoms and O for oxygen. The lower case n represents the number of atoms. The OH at the end of the formula is called the "hydroxyl" group and is what makes alcohols different from other hydrocarbons.

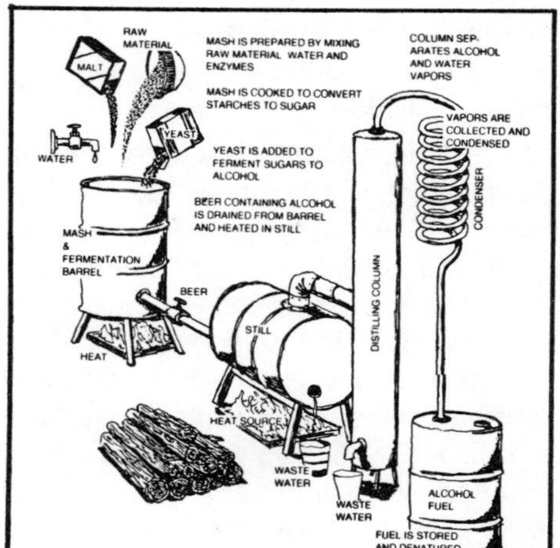

Fig. 2-1. The basic process of alcohol production.

It's not important for you to memorize these formulas, but only to know that basic differences exist. For example, the chemical formula for methanol is:

$$CH_3OH$$

It's called methanol because it is a combination of a methyl (CH_3^+) and hydroxyl (–OH) radical. A radical is what chemists call an atom or group of atoms that want to share their electrons with another radical.

Methanol alcohol is included here because it is the "other" alcohol most often discussed when talking about alcohol fuels. Methanol and ethanol are quite similar chemically, but ethanol has a higher Btu (heat) content which makes it a more attractive motor fuel. As mentioned earlier, ethanol is the easiest alcohol for the do-it-yourselfer to manufacture.

The chemical formula for ethanol alcohol can be written one of two ways:

$$C_2H_5OH \text{ or } CH_3CH_2OH$$

Both are correct, but the one on the right is more descriptive of the molecular arrangement of atoms.

With ethanol, the two basic components that combine to make the alcohol molecule are an ethyl ($C_2H_5^+$) and hydroxyl (–OH) radical. So much for the chemistry lesson.

Making Alcohol

The basic process (Fig. 2-1) of making alcohol from any biomass raw material involves three fundamental steps:

—Mash preparation.
—Fermentation.
—Distillation.

The general idea is to take a raw material, grind it up and treat it in such a way that, when mixed with water, the resulting "mash" will contain about 10 percent to 15 percent simple sugar. Preparing the mash sounds easy enough, but in actual practice it involves a number of steps that are necessary to free the sugar from the raw material. It all depends on the raw material used, but generally mash preparation includes cooking, the addition of acids and the addition of substances known as "enzymes." All this will be explained in much greater detail further on. For now, just remember that mash preparation involves breaking down the raw material to free simple sugars into solution.

The next step is the addition of yeast to the mash. Yeast are microscopic fungi that feed on simple sugars. The byproducts of their metabolism are carbon dioxide gas and ethanol alcohol.

$$Sugar + Yeast + CO_2 + Ethanol = more\ Yeast$$

This is the process of fermentation. As the yeast grow and multiply in the mash, the escaping CO_2 gas causes the brew to bubble and foam. This is a good indication that the yeast are thriving. After a few days, the bubbling activity will die down. At this point, the mash will contain about 8 percent to 12 percent ethanol and will be ready to harvest.

Fermentation dies down after a few days because the yeast have consumed most of the available sugar and the rising concentration of alcohol in the mash hinders further growth. Since the yield of alcohol has reached its peak, the liquid that contains the alcohol—called the *beer*—can now be separated from the rest of the mash.

The beer is drained off or strained from the mash. The leftover solid residue from the mash still contains most of the basic ingredients of the original raw material (minus the sugars). For corn or other grains, this includes vitamins, minerals and proteins. This nutrient rich residue can be fed wet to livestock, or dried and used as a high-protein food supplement.

The beer can now be poured into a "still" so that the alcohol can be separated or *distilled* from the liquid—which is mostly water. This is the most difficult and energy intensive part of making alcohol because water and alcohol are not easy to separate.

Since there is no easy way to separate the alcohol and the water by straining or filtering, you must take advantage of a physical difference between the two. Water boils at 212 degrees F and ethanol boils at 173 degrees F. By heating the still to just over 173 degrees F, more alcohol vapor will boil off than water. As the vapors rise, they are carried off through a pipe to a *coil* where the vapor is cooled and condensed into liquid again. By repeating the process of vaporization and condensation, the concentration or proof of the alcohol can be increased up to a maximum of about 90 percent to 95 percent pure ethanol or about 180 proof to 190 proof (50 percent alcohol equals 100 proof).

Removing those last traces of water becomes extremely difficult. Once the proof reaches about 195, it becomes physically impossible to separate the remaining water by distillation alone. To get those last few drops of water out, *drying agents* or chemicals such as benzene or calcium oxide (lime) must be used. When mixed with the liquid, these chemicals react with the water but not the alcohol. With lime, for example, the water reacts with the calcium oxide to form an insoluble compound called calcium hydroxide. The sediment settles to the bottom of the mixing tank and the *anhydrous* (free from water) or 200 proof alcohol is drained off and filtered.

Fortunately, you don't have to worry about small amounts of water in your alcohol fuel if you're burning straight alcohol. In fact, the presence of a small amount of water actually helps the fuel burn better. Alcohol fuels with proof as low as 140 (70 percent ethanol, 30 percent water) have been successfully used in engines with modified carburetors. Fuel as low as 100 proof (50 percent ethanol, 50 percent water) has been burned in vapor injection systems. Most simple conversions, however, seem to work best with alcohol fuels in the 180 proof range (10 percent water content).

The only instance where anyhdrous (or 200 proof alcohol) is required is when alcohol is blended on a commercial basis with gasoline to make gasohol. Because gas and water don't mix very well, the presence of more than a very small percentage of water in gasohol might cause the mixture to separate on a cold morning. This could cause hard starting or stalling if straight water is drawn into the carburetor. Certain chemicals such as benzene can prevent phase separation in gasohol blends with a higher water content, but most such chemicals are either too expensive or too toxic for the do-it-yourselfer to use.

Mash Preparation

Now that you've been introduced to the rudiments of making alcohol fuel, let's go back and take a closer look at some of the specifics of making fuel from corn.

- Grind or mill the corn kernels through a three-sixteenth inch screen. Consistency should be sand-like but not as fine as flour.
- Add about 16 gallons of water per bushel of corn and mix thoroughly.
- Pre-malt the mash by adding 10 percent of the total amount of malt or enzyme to be used. The total amount of barley malt should be 6 to 8 pounds per

bushel of corn. If commercial enzymes are used instead, use only about one-half pound per bushel.

- Cook the mash at a slow boil for about 30 minutes with constant agitation. This prepares the starch for conversion to fermentable sugar.
- Cool the mash to around 145 degrees F and dilute the mash by adding an additional 16 to 18 gallons of water per bushel of corn. The water will help cool the mash.
- Add the remaining malt or enzymes and mix thoroughly. Continue agitation for 30 minutes. The conversion of starch to fermentable sugar should now be complete.
- Adjust the pH to a level of from around 4.5 to 5.0 for optimum yeast growth.

Fermentation

- Add brewer's yeast in a proportion of 2 to 4 ounces per bushel of corn.
- Allow the mash to ferment for 72 hours at 85 degrees F to 90 degrees F. When bubbling ceases, fermentation has ended. The yeast will have converted most of the sugar into ethanol.
- Separate the liquid beer (which will be about 10 percent alcohol, 90 percent water) from the mash by straining. The mash residue can be fed wet to livestock or dried for later use as a high protein food supplement for animals.

Distillation

- Heat the beer at 175 degrees F to 180 degrees F in the still. This will vaporize the alcohol and leave most of the water behind.
- Route the alcohol vapors through a reflux column to remove more water and increase the concentration of alcohol.
- Condense the vapors as they leave the column. The resulting proof will depend on the efficiency of the column.
- Denature the alcohol fuel in accordance with the law.

Commercial-scale Fuel Production

The following describes the conversion process used by the Archer Daniels Midland alcohol production plant near Decatur, Illinois. The process yields ethanol alcohol, corn germ meal, corn oil and high protein gluten. Obviously, not all 13 steps are required if simple fuel production is the goal. Steps 3 through 7 would be omitted and step 12 would not be necessary if you were using straight alcohol fuel in a converted engine.

1. Steeping to loosen various parts of the corn kernel.
2. Grinding to break the corn kernel apart.
3. Germ separation to recover the corn oil and corn germ meal.
4. Fine grinding to separate the starch and gluten fractions.
5. Removal of fiber for corn gluten feed which is 21 percent protein.
6. Removal of protein for corn gluten meal which is 60 percent protein.
7. Concentrating the starch by removing excess water.
8. Cooking the starch in the presence of the liquifying enzymes which convert the starch chains to dextrins.
9. Cooling and adding saccharifying enzymes to convert the dextrins to glucose.
10. Yeast is then added to convert the glucose to ethanol alcohol and carbon dioxide.
11. The alcohol (10 percent in the fermenter) is distilled to a concentration of 96 percent to 96.5 percent pure.
12. The purified alcohol is then redistilled to 99.75 percent to 100 percent pure.
13. The anhydrous alcohol is denatured in accordance with the law and sold for blending with gasoline to make gasohol.

SUGAR SOURCES

Because yeast need a supply of simple sugar in solution to make alcohol, raw materials are needed that can supply sugar in sufficient quantities to make the process worthwhile. The most abundant source for this sugar is plant matter.

Plants are mostly water, carbohydrate and protein—with differing amounts of these ingredients concentrated into different portions of the plant. It's the carbohydrates we're interested in. Carbohydrates fall into one of three basic categories:

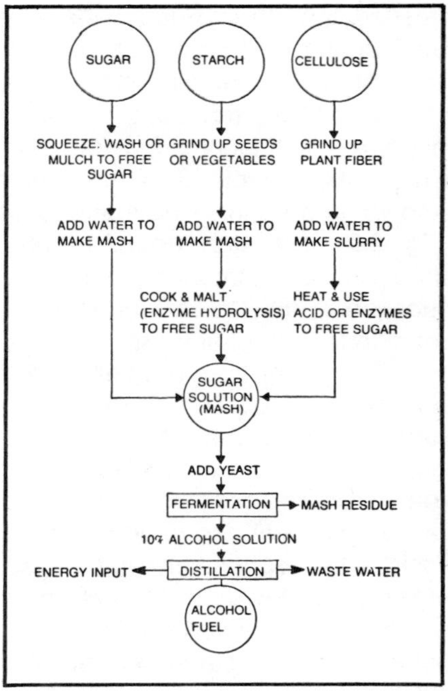

Fig. 2-2. A flowchart for raw materials.

Sugars. Glucose, fructose and maltose are found in large concentrations in such plants and plant products as sugar beets, sugar cane, cane sorgum, molasses, fresh or dried fruits, whey and skim milk.

Starches. These are larger, more complex molecules formed from groups of simple sugars joined together. Starches include potatoes, sweet potatoes, corn, wheat, milo, grain sorgum, barley, rye, artichokes, cacti, manioc, arrowroot and pumpkins.

Cellulose. Composed of long chains of sugars and lignin, cellulose is the basic building block of all plant fiber. It is found in such things as corn stalks, corn cobs, hay, straw, grass and wood. Cellulose is also the primary ingredient in paper (from wood pulp) and manure (from undigested plant fiber).

Although nearly any crop or plant residue can be used in the manufacture of alcohol fuel, some provide far greater yields than others or are easier to process. The easiest type of materials to make alcohol from are the high sugar content group. This includes sugar cane, sugar beets and molasses. To free the fermentable sugars, the raw material need only be mulched,

33

pressed or washed. Sugar cane is currently the leading alcohol fuel feedstock in Brazil because it's fairly simple to process and it is that nation's leading crop.

The starchy materials in group two (corn, grain, etc.) also make excellent alcohol fuel feedstocks since they can produce a high alcohol yield. Starches, however, require an additional step when preparing the mash to free the fermentable sugars (Fig. 2-2). The starch must be *malted* to break it down into simple sugars. Starches are popular feedstocks in this country because of their great abundance and suitability to growing conditions in the Midwest.

The higher fiber materials (stalks, cobs, paper, etc.) can also be used, but the breaking down of cellulose into simple sugar is a more difficult process. Such materials are therefore less attractive than sugar or starchy feedstocks for the do-it-yourselfer. The conversion of cellulose to a fermentable sugar can be practical and economical when done on a large-scale basis. A tremendous amount of research has been done in this area the past few years to develop schemes for recycling municipal garbage or scrap paper into alcohol fuel. To illustrate the potential, a ton of paper will yield about half a ton of fermentable sugar.

MAKING FERMENTABLE SUGAR

Starches and cellulose can be broken down by acid hydrolysis or by enzyme hydrolysis.

With acid hydrolysis, the feedstock material is ground up and mixed with a dilute acid solution to form a slurry. The acid then attacks the chemical bonds that hold the starch or cellulose molecules together, breaking the molecules apart and thereby freeing the simple sugars into solution. The sugar can then be fermented and made into alcohol.

In commercial applications, the raw feedstock is ground up and mixed with water in a ratio of about 10 parts liquid to 1 part solid. The slurry is then mixed with an 8 percent to 10 percent weak acid solution and cooked under pressure at about 360 degrees F to 440 degrees F. The result, in the case of scrap paper, is a conversion of about 80 percent of the cellulose into fermentable sugar. This process has proven itself to be practical and economical in large-scale applications where large volumes of materials that contain cellulose such as garbage, paper or wood pulp must be broken down and processed quickly. Unfortunately, the technology involved and the economics of scale limit acid hydrolysis to the realm of the large-scale venture. Because of this, I will forego the details of acid hydrolysis and focus instead on enzyme hydrolysis.

Enzyme hydrolysis, or "malting" as it is often called, is the process mankind has used for thousands of years to break down starchy feedstocks to make fermentable sugars. With this process, enzymes from barley malt are mixed with the ground up starch and water. The enzymes then do approximately the same thing as the acid and attack the chemical bonds that hold the starch molecules together. This frees the simple sugars so that they can be fermented to make alcohol.

Specific enzymes must be used to attack specific types of molecules. Fortunately, barley malt contains an abundant supply of a common type of enzyme, alpha amylase, that works quite well in breaking down starch in a variety of crops, including corn and wheat.

Enzymes are also showing great promise for breaking down cellulose. Cellulose is difficult to decompose because its molecules can consist of chains of simple sugars that are up to 3000 sections long. The chains curl and twist in such a way that they resist chemical attack. That's why wood makes a great building material. However, wood will rot because certain fungi and bacterial possess the right type of enzymes to digest cellulose.

ENZYMES

Enzymes are biological catalysts. They are a special group of substances that living organisms use to break down other chemicals or speed up chemical reactions. Enzymes are necessary to break down food for digestion, for muscle movement, for nerve reactions and to carry on most of the basic functions of life. Saliva, for example, contains alpha amylase to help convert starch into sugar before it reaches your stomach. Chew on a piece of starch awhile and you'll begin to taste the sweetness of the sugar being released.

To better understand enzymes, consider the biology of a typical seed. A seed contains a new plant embryo and a supply of "stored" energy in the form of starch to help get the new plant started. Split a corn seed open and you will see the little embryo surrounded by the rest of the seed, which is mostly starch. For the seedling to sprout and grow, enzymes must first be released within the seed to break the starch down into simple sugars the embryo can metabolize. Until these enzymes are released, nothing will happen. The stored energy remains locked in the starch and seed lies dormant.

The best source of enzymes, therefore, is sprouting seeds. One type of plant that produces an abundance of enzymes in its sprouts is barley. You can learn how to sprout barley in a later chapter, so for now just remember that barley is probably the most common source of enzymes.

The sprouting barley, also called *malt*, is chopped up and added to the ground starch solution to "hydrolyze" or decompose the starch into sugar. This is where the term "enzyme hydrolysis" comes from. If the term sounds too technical, just call the process *malting* as brewers have for hundreds of years.

Enzymes can be home-grown to save expense or purchased from commercial suppliers in a concentrated form. The latter is often preferred since specific enzymes can be obtained for a minimum of effort. The added convenience is often worth the extra cost. A list of suppliers is included in the appendix.

Enzymes, pH and Yeast Growth

Because malting and fermentation are essentially biological processes, it's important that mash be properly pH balanced to create a favorable environment for enzyme action and yeast growth. Maintaining proper pH control is also the best protection against bacterial contamination of the mash, which would seriously reduce the alcohol yield of the batch.

What is pH? It's a measure of the relative acidity of the mash solution. It is based on a scale of 1 to 14, with 1 being highly acidic, 7 neutral and 14 highly alkaline. Pure water, for example, is neutral or 7.

Water is H_2O, or two hydrogen atoms and a single oxygen atom. When a water molecule breaks apart, it tends to split into a free hydrogen atom (H^+) and a paired hydrogen and oxygen atom $(-OH)$. The plus and minus signs mean the hydrogen and hydroxyl want to combine again to share their electrons. If an even number of both are not present, then the remaining one will try to combine with other molecules or radicals that might be present. What has all this got to do with acids and alkalines? Plenty, since acids create an excess of H^+ and alkalines create an excess of $-OH$. Add hydrochloric acid (HCl) or sulfuric acid (H_2SO_4) to the mash, and the concentration òf H^+ rises, increasing the acidity of the solution.

By definition, pH means the "percentage of hydrogen." It is

a measure of the concentration of hydrogen. For math buffs, the numbers 1 through 14 refer to the negative logarithm of the proportion of hydrogen to hydroxide. Understanding the chemistry isn't as important as remembering that the lower the pH number, the more acidic the mash.

Test paper or a pH meter can be used to measure the pH of the mash. A strip of test paper is dipped into the solution and changes color. The color is then compared to a reference chart that comes with the test paper. Test papers are available in different ranges, but a range of 4 to 8 or 3 to 9 is most useful when making alcohol. The papers are available through laboratory supply companies, swimming pool supply houses and some garden shops.

The pH meter tests the pH by running a small electric current through the solution. It's quick and exact, but also more expensive. Sophisticated laboratory models can be purchased from laboratory or chemical supply houses, but less expensive "plant pH" meters will work just as well if they offer the proper pH range on their scale. These can be purchased for less than $10 at many garden or plant shops.

Controlling pH during the malting process and fermentation is important for three reasons. Enzymes need a certain pH range to do their job properly, yeast will only grown in a slightly acid solution, and bacterial contamination is less likely in an acidic mash.

The greatest threat of bacterial contamination occurs after the mash has been cooked. If airborne bacteria get into the mash after cooking or during fermentation, they will compete with the yeast for the available nutrients in the mash and lower the yield of alcohol by producing lactic acid. Contamination can be prevented by any or all of the following:

- Cleaning the cooking and fermentation tanks between runs.
- Reducing the cooling period between cooking and fermentation to a minimum.
- Sealing the cooking and fermentation vessels with air or "fermentation locks" to prevent outside contamination.
- Using penicillin or other antibiotics to suppress undesirable organisms.
- Maintaining a pH balance that encourages yeast growth but suppresses other organisms.

Yeast prefer a slightly acidic solution, with a range of 4.5 to 5.0. This severely limits the growth of other organisms. If the pH is maintained below 5.0 during fermentation, competing organisms won't have a chance to grow. Above 5.0, bacterial growth can run rampant. The pH range should also not be maintained too low, as anything much less than 4.4 will inhibit the sugar to alcohol conversion process.

The optimum pH range for yeast growth will depend on the strain of yeast being used, but most will fall in the range of 4.8 to 5.0.

How do you adjust the pH of the mash? By adding small amounts of acid or alkaline (such as sulfuric acid, etc.). Most mashes will need the addition of some acid, as most grain mashes have a natural pH of 5.4 to 5.6. Other feedstocks such as the high sugar content substances (molasses, fruits, etc.) have a naturally alkaline pH and must be adjusted prior to fermentation.

When adjusting the pH of the mash, it is very important not to add to much acid—a little bit goes a long way. After checking the pH of the mash, some acid should be diluted with water and thoroughly mixed with the mash being checking the pH again. If it's still too high, add a little more acid. The idea is to lower the pH gradually (increase the acidity) so as not to exceed the optimum pH for yeast growth. If you find that you've added too much acid, the pH can be adjusted the other way by adding a small amount of alkaline such as sodium hydroxide (caustic soda). This will increase the pH. By "fine tuning" the mash, you should be able to maintain maximum yeast growth and alcohol yield.

The pH of the mash can also be adjusted by adding a small amount of mash residue left over from a previous run. This is called *backslopping*. By adding 10 percent to 20 percent stillage residue to the mash prior to fermentation, the acids in the stillage will help lower the pH to the desired level. The stillage contains additional nutrients that will aid yeast growth. Backslopping also makes the mash more tolerant of changes in pH.

Cleanliness

Cleanliness is essential to prevent the growth of bacterial in mash residues that collect in tanks, pipes and stills. Any ordinary disinfectant can be used to clean the equipment between runs, but formaldehyde or ammonia are probably the most effective. After cleaning, the equipment should be washed with clean water to remove any traces of cleaning chemicals that might inhibit yeast growth.

Chapter 3
Feedstocks

The object of do-it-yourself fuel production is to find a cheap, readily available feedstock that can be made into fuel with a minimum of effort. Three things should be considered:

—The availability and cost of the raw material.
—The relative yield of alcohol from the raw material.
—The degree of difficulty to process the raw material into alcohol.

Sugar feedstocks are the easiest to process, with starches being second in line. Cellulose, on the other hand, is more difficult to break down. In most instances the final choice will probably be determined by the relative abundance of a particular feedstock in your area.

If you're a farmer in Iowa, Illinois or Indiana, for example, corn would be a logical choice. Corn gives a high alcohol yield, it grows well given the climate conditions of the area and the techniques for planting and harvesting it are already well established. Since corn is the number one crop in the area, it only seems logical to use some of the surplus to make alcohol. Using the existing crop means no additional money will have to be spent for different harvesting equipment.

This is not to say corn is the best choice for everyone. In Nebraska and the Dakotas, wheat is the natural choice. In Hawaii, the best crop might be sugar cane. In Southern California or Florida, citrus fruits or leftovers from a fruit canning operation might be the best choice. In Idaho, it's potatoes. In

Table 3-1. Alcohol Yields.

Feedstock	gallons/ton
whey, dry	95
Wheat	85
Corn	84
Buckwheat	83
Raisins	81
Grain sorgum	80
Rice, rough	80
Barley	79
Dates, dry	79
Rye	79
Prunes, dry	72
Molasses, blackstrap	70
Cane sorgum	70
Oats	64
Cellulose (pure)	62
Figs, dry	59
Wood & agricultural residue	47
Sweet potatoes	34
Yams	27
Potatoes	23
Sugar beets	22
Figs, fresh	21
Jerusalem artichokes	20
Pineapples	16
Sugar cane	15
Grapes	15
Apples	14
Apricots	14
Pears	12
Peaches	11
Plums	11
Carrots	10

certain areas of the South, rice is in plentiful supply. In forested regions such as the Pacific Northwest, wood is the number one commodity. In metropolitan areas, it's garbage and scrap paper. At a cheese processing plant in Wisconsin, it might be whey. As you see, the best feedstock depends on the circumstances and alternatives available at the time. Making a selection means considering the alternatives, the costs, the degree of difficulty involved in planting, harvesting and processing a given feedstock.

For scrap paper, garbage, manufacturing wastes and spoiled grain or sprouted potatoes, the costs of the raw material might be very low because the problem of disposal is very high. Often you will find that the raw material is free for the asking. In some instances, you might actually be paid to haul it away. In the case of the plant owner, you might have to pay to have it hauled away. Regardless of which side of the garbage bin you're on, recycling such materials into a useful energy resource not only solves the problem of waste disposal but also creates a valuable service for society.

In addition to costs availability, the yield of alcohol, per ton of raw material is worth considering. Wherever possible, the higher yielding feedstock should be selected if two or more alternatives are available.

The figures in Table 3-1 were compiled over 50 years ago by researchers investigating the potentials of different alcohol feedstocks. The yields are based on the average fermentable sugar content that can be converted into alcohol.

Although these figures are approximate values, they can give you a pretty fair idea of how the various feedstocks compare. These numbers won't vary much more than 10 percent either way.

The actual amount of alcohol fuel that can be obtained from a given feedstock will depend on:

—The proof of your final fuel (based on pure or 200 proof ethanol).
—How well the yeast performed during fermentation.
—Whether or not the mash was contaminated.
—How thoroughly the beer that contains the alcohol was separated from the mash after fermentation.
—How efficiently the distilling equipment separated the water and alcohol.

FIGURING ENERGY POTENTIAL

You might be surprised at the amount of fuel that can be extracted from a given crop. Although thinking about crop yields in terms of gallons-per-ton might sound strange, Table 3-1 is organized so that you can see the relative yields of the various feedstocks. You can convert the gallons/ton figures to gallons/ bushel by determining how much a bushel of corn or wheat weighs and then doing the math or you can use Table 3-2.

Table 3-2. Potential Fuel Yields.

Corn	2.6 gallons/bushel
Grain sorgum	2.6 gallons/bushel
Wheat	2.7 gallons/bushel
Barley	2.1 gallons/bushel
Potatoes	1.2 gallons/100 lbs.

42

Table 3-3. Fermentable Sugar Content.

Grapes	15.0%	Peaches	7.6%
Bananas	13.8%	Oranges	5.4%
Apples	12.2%	Prickly pear	4.2%
Pineapples	11.7%	Watermelon	2.5%
Pears	10.0%	Tomatoes	2.0%

Again, these figures are averages and are based on pure (200 proof) ethanol. You can use them to estimate the potential yield for a particular feedstock. For example, to figure how many gallons of alcohol you could produce from an acre of corn, multiply your average crop yield in bushels-per-acre times the number of gallons-per-bushel given for corn.

(crop yield in bu/acre) × (gallons/bushel) = gallons fuel/acre

If you average 80 bushels of corn per acre, then 80×2.6 equals 208 gallons/acre. This means every acre of corn on your farm can yield up to 208 gallons of ethanol. If you're burning 180 proof fuel (10 percent water) in your converted tractor, then the actual amount of fuel per acre would be 10 percent greater because you're leaving 10 percent of the water in the fuel. The total is now 229 gallons of alcohol fuel per acre.

Remember, leaving a small amount of water in the alcohol effectively extends the fuel supply, just as adding a small amount of alcohol to gasoline (to make gasohol) effectively extends the gasoline supply.

FEEDSTOCK CHOICES

The ingredients for alcohol fuel production fall into one of three basic categories: sugars, starches or cellulose. Sugars are the easiest to process, starches are next easiest and cellulose is fairly difficult.

Sugars, also called *saccharine materials*, are easiest to convert to alcohol because the sugars can be extracted and fermented with a minimum of effort. By mulching, pressing or washing, most of the fermentable sugar in the raw material can be recovered. Enzymes are not required since there are no starches to be broken down.

As with other feedstocks, the alcohol yield for saccharine materials depends on the fermentable sugar content of the original crop. Table 3-3 shows some averages for common crops. Depending on how much of the sugar you can extract or squeeze out of the feedstock, the total amount of alcohol produced will vary with the amount of sugar available. Normally, 75 percent to 80 percent of the sugar can be extracted.

Obviously, some sugar feedstocks are much better than others as a raw material for making fuel (Table 3-4). Sugar cane and molasses, for example, can give fairly high yields. Watermelons are mostly water and squeezing a ton of them would only give you enough fermentable sugar to make about 3 gallons of fuel. That is hardly worth the effort. The same goes for apples. If you squeeze 75 percent of the sugar out of a ton of apples, you get enough sugar to make about 12 or 13 gallons of fuel. Needless to say, the much higher yields offered by corn, wheat and other grains makes many of the starchy feedstocks appear far more attractive—unless you've got a lot of apples or watermelons you want to get rid of.

Although the starchy feedstocks offer higher yields, they are somewhat more difficult to process because cooking and malting are necessary to break down the starches into fermentable sugars. Here's a sampling of average starch and sugar contents for some typical grains:

| Barley | 50% | Rye | 59% | Oats | 50% |
| Maize | 66% | Sorgum seed | 67% | Wheat | 65% |

The average alcohol yield per ton will depend on how thoroughly the starches are converted to sugar prior to fermentation. If the cooking and malting steps of mash preparation are not completed, the yield will be reduced.

Table 3-4. Raw Material Comparisons.

Feedstock	Sugar	Starch	Cellulose
	sugar cane, fruits, etc.	corn, grain, potatoes, etc.	plant residue, garbage, paper, etc.
Yield (gallons/ton)	medium to low (except for whey which is high)	high to medium	medium
Degree of difficulty to process	simple	relatively easy	difficult for do-it-yourselfer
Conversion process	direct fermentation of sugar	cooking and malting needed to free sugar	enzyme or acid hydrolysis needed to free sugar
Cost of feedstock	moderate	moderate	cheap or free
Source of feedstock	main crop crop surplus damaged crop crop wastes	main crop crop surplus damaged crop	municipal garbage agricultural residue wood waste (Most abundant feedstock)

Cellulose materials are the most difficult to break down into fermentable sugars because of the chemical nature of cellulose resists decomposition. Much research is being done to develop improved technologies for converting cellulose into sugar.

CELLULOSE

The main advantages cellulose offers as an alcohol feedstock are that it is inexpensive and it is one of the most abundant resources on earth. It's the basic structural fiber of all plant life. Cellulose is especially plentiful in plants such as trees, woody shrubs, straw, hay, grass, wood wastes and agricultural crop residues. Paper is almost pure cellulose and municipal garbage is typically 35 percent to 55 percent paper.

The main problem has been finding an economical way to convert cellulose to fermentable sugar. The conversion can be done using either acid or enzyme hydrolysis, but the economies involved seem to be more favorable for the large-scale rather than small-scale fuel producer. For the do-it-yourselfer, it makes more sense to burn cobs, stalks, wood chips or paper to heat a still (or make electricity) than to use them as the alcohol feedstock. Another problem with cellulose for the do-it-yourselfer is the sheer bulk involved. A ton of corn stalks is a lot of stalks to cut, haul and grind into a slurry. If all the time and effort were taken into account, it would hardly be an energy efficient undertaking. The few gallons of fuel most cellulose residues would produce would not be worth the effort.

Recycling scrap paper, municipal garbage or lumber wastes is another story. Here the situation is more favorable because the wastes are already concentrated in one location (a saw mill, a municipal landfill, a printing factory, etc.). Waste disposal has created serious problems for many businesses and communities and recycling the wastes into fuel would appear to be the ideal solution.

Although most cellulose-to-sugar technologies are based on acid hydrolysis, much progress has been made in developing enzymes that will attack cellulose molecules directly. The enzyme that does this is called *Cellulase*. This enzyme is used by numerous strains of fungi, algae and bacteria to decompose plant cellulose. The problem has been isolating an organism that produces it in sufficient quantities to be economical.

The U.S. Army Laboratories at Natick, Massachusetts has been pioneering research in this field. A strain of fungus or mold that caused canvas tents and clothing to rot in the tropics was isolated and identified by researchers. They called this strain of cellulase producing fungi *Trichoderma viride*. After considerable effort, the scientists managed to develop a mutant strain that produced four times the Cellulase enzyme as its wild counterpart. The enzyme has produced reasonably good sugar yields from pure cellulose (such as scrap paper). Research is now focusing on improving the yields of forest and agricultural residues that contain lignin.

FUTURE TRENDS

According to a report issued by the U.S. Department of Energy, the biomass resources annually available in the United States for alcohol fuel production currently totals 800 million dry tons. Wood comprises 61 percent of the total, agricultural residues 23 percent, municipal solid wastes 10 percent, grains 5 percent and food processing wastes 1 percent. The 800 million tons does *not* include the crops used for food, animal feed or export demand.

The report goes on to say if a national sweet sorgum program were developed and promoted, grains and sugar crops could account for 15 percent of the surplus biomass resource base by the year 2000.

How much fuel does this translate to? Based on current technology,—conversion of only the food processing wastes, surplus grains and sugar crops could yield 12.2 billion gallons of ethanol by the year 2000. If you add ethanol produced from municipal solid wastes, that figure would nearly double. Add to that the ethanol that could be produced by recycling agricultural residues and wood wastes and you have quite a total.

Chapter 4
Making Mash

Once you've chosen the raw material, you're ready to begin the process of making alcohol fuel. The first and most important step towards your goal is the preparation of the mash.

What exactly is the mash? Think of it as a kind of soup. The primary ingredients in this soup will be water and the components that contain sugar (sugar, starch or cellulose) from the base feedstock. Your job as chef is to make sure these ingredients are mixed in the right proportions so that the end product will be suitable for fermentation. Remember, yeast need *dissolved* sugar to make alcohol. The only place they'll get it is from your particular feedstock. The idea, therefore, is to make your soup as palatable as possible to the yeast. This means the right balance of ingredients (sugar and other nutrients), proper pH (4.5 to 5.0) and a temperature that will promote rapid yeast growth (around 85 degrees F).

SUGAR CONCENTRATION

The most critical factor in preparing the mash is the sugar concentration. For optimum conversion of sugar to alcohol, the mash should have a dissolved sugar content greater than 10 percent but less than 20 percent. The ideal percentage is about 15 percent. If the mash contains less than 10 percent dissolved sugar, the yeast will do poorly because of a lack of sufficient food. This will limit the alcohol yield of the batch. If there's more than 20 percent sugar, the yeast might go wild. The frenzied activity produces lots of yeast but little alcohol. Since most of the energy goes into reproduction. It's also quite possible that some strains of yeast might suffer toxic effects from sugar concentrations over 20 percent. Dead yeast doesn't make alcohol. It's best, therefore, to aim for the 15 percent sugar content recommended.

MAKING MASH FROM A SUGAR FEEDSTOCK

As mentioned in Chapter 2, sugar feedstocks are the easiest to work with, even though the yields are not as high as those that can be obtained from starch or cellulose feedstocks. The advantage with sugar feedstocks is that no cooking or malting is required. The sugar is already in a fermentable condition. All that need be done is to extract it from the particular crop.

If apples are the raw material, the mash is prepared by pressing the juice that contains sugar from the fruit. Since the juice will be mostly water, the sugar content should be well under the 20 percent limit. If the sugar content is under 10 percent, excess water can be removed by evaporation or boiling. Evaporation is the preferred choice since it requires no energy input on your part. Simply pour the juice into a shallow trough or pan and let solar energy take care of the rest. As soon as enough water has evaporated to increase the sugar content above 10 percent, remove the juice, check the pH (add acid if too high) and add yeast to begin fermentation.

Sugar syrups such as molasses have sugar contents as high as 50 percent. This is much too sweet for the yeast to handle. Here, water *must* be added to dilute the mash down to the desired 15 percent range.

With crops such as sugar cane, the sugar is removed by first mulching the crop and then washing with water. The washing should be repeated several times to dissolve as much of the sugar as possible. After about three washings, however, a point of diminishing returns is reached and the concentration of dissolved sugar obtained from the crop will be too low to be worthwhile. Over dilution of the mash should also be avoided by using a minimum amount of water during each washing.

MAKING MASH FROM A STARCH FEEDSTOCK

With starch feedstocks, mash preparation is more complicated. The starch material must first be ground (mashed) and mixed with water to form a thick slurry—about 30 percent to 35 percent suspended solids. The ratio of liquid to solids should be about 2 parts water to each 1 part ground grain. The starch must then be broken down into fermentable sugar by enzymes or acid. The starch-to-sugar conversion process is known as *malting*.

Despite these additional steps, starchy feedstocks are quite popular because of their relative abundance in this country and the high yields of alcohol they produce.

The first step in the mash making process is to grind the raw material to a sand-like consistency—but not as fine as flour. If it's too fine, the mash might thicken to the point where it becomes difficult to stir and handle.

The reason for milling starch is twofold. Starch does not dissolve readily in water so it must be ground. The grinding also exposes maximum surface area to the enzymes that will be added later. This makes the conversion of the starch to fermentable sugar possible.

Cooking

The next step in preparing the mash is to boil the soup. Cooking accomplishes several things. It helps to sterilize the mash to eliminate organisms that might interfere with the yeast during fermentation. It also helps to swell or "gelatinize" the starch molecules so they can be more easily decomposed into their simple sugar building blocks.

Cooking is usually done at a slow boil for about 30 minutes. Although you'll be using a simple batch process to prepare your mash, most commercial alcohol plants use pressure cooking on a continuous basis to minimize production time. Increasing the pressure means you can use a higher temperature for a shorter period of time. A higher temperature is also more likely to produce a 100 percent sterilization of the mash, since some organisms can survive boiling at atmospheric pressure for a short period of time.

To illustrate how pressure cooking speeds up the production process, 15 pounds of pressure would enable you to cook the mash at 250 degrees F in 15 minutes. Going as high as 160 psi, you could cook the mash at 360 degrees F in 30 seconds. Pressure cooking, however, requires heavy pressure vessels which are too expensive and dangerous for the do-it-yourselfer. You will get the same results with the tried and proven batch technique of cooking the mash in an open vat.

Always use the cheapest source of heat that is available in order to hold your fuel production costs to a bare minimum. Use firewood, wood scraps, trash, corn cobs or methane from compost or manure. Remember that thick mash takes less energy to heat than thin mash. Excess water only eats up energy when it comes to cooking.

After the mash has been cooked, more water must be added to cool and dilute the brew. Because the cooked mash is too concentrated for proper malting, more water is needed. Add another 2 parts water for each 1 part solids already in the mash. Depending on how much water was lost during the boiling, you might want to add a little extra water. For example, if the recipe calls for an additional 16 gallons of water per bushel of grain, you might want to increase this to 17 to 18 gallons per bushel to make up for the lost water.

Malting

After the mash has been cooked, barley malt or commercial enzymes are added to break the swollen starch molecules into fermentable sugar.

The amount of malt used should equal about 10 percent to 15 percent of the weight of the solids used to prepare the mash. For instance, you should use about 5.6 to 8.4 pounds of malt for each bushel of corn, if corn is your raw material. That's because corn weighs about 56 pounds per bushel. Some average weights for common starch feedstocks are given in Table 4-1.

Commercial enzymes such as Alpha Amylase, available under the trade names Taka-Therm, Tenase and Termamyl 60, are more concentrated than barley malt. Therefore, they are used in much smaller dosages. A rule of thumb is to use an amount of enzyme equal to 1 percent of the weight of the solids in the mash. For a bushel of corn, that would be a 1 percent of 56 pounds, or about one-half pound of enzyme.

Table 4-1. Average Weights of Common Starch Feedstocks.

Starch Feedstock	lbs./bushel
Wheat, beans, peas, potatoes	60
Rye, corn, mixed grain	56
Barley, buckwheat	48
Rough rice	45
Oats	32

Malt or Enzyme

Both barley malt and commercial enzymes work equally well and each has its own particular advantages. Enzymes are more expensive but are considered to be easy and convenient to use. Barley can be purchased in a dried, ground form or grown at home. Sprouting barley malt, however, takes some time and effort so it might not be for everybody.

Sprouting barley involves soaking barley seeds in water for 8 to 12 hours. The water should be changed several times and kept at room temperature. The seeds are then spread out in a moist, dark environment and left to sprout. After several days, the sprouts should appear. As soon as they are about one-half of an inch long, they should be gathered and ground. The "green" malt must then be used immediately to prevent spoilage. Malt can be stored indefinitely if it is first dried and then sealed in an airtight container. This is how commercially available malt is prepared.

When malting, the mash should be stirred constantly to keep the enzymes circulating and interacting with the starch. Stirring should be provided by an electrically powered paddle unless you've got strong arms. Once the process is underway, it should take about 30 minutes or so to complete. During this time, the temperature of the mash should be maintained at approximately 145 degrees F. The enzymatic process is temperature dependent. At 145 degrees F, the starch-to-sugar conversion proceeds at peak efficiency. Be warned, however, that temperatures above approximately 170 degrees F to 175 degrees F will destroy the sugar making ability of the malt. Some commercial enzymes will work at higher temperatures. Unless a higher temperature is recommended by the supplier, stick wtih 145 degrees F.

Iodine Solution

To determine whether or not all of the starch has been converted to sugar, an iodine test can be used. Iodine reacts with starch to form a blue to black coloration, but it does not react with sugar. To test the mash, remove a small sample of liquid and add a few drops of iodine. If nothing happens, the malting process is complete and all the available starch has been converted to fermentable sugar. If the drops cause a blue to black coloration to appear, the malting process isn't complete. The mash needs more time or more enzymes. The darker

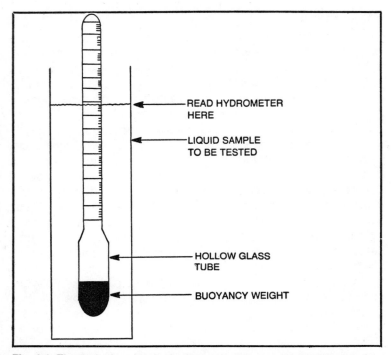

Fig. 4-1. The typical parts of a hydrometer. The buoyancy weight at the bottom determines where the hydrometer will float in a liquid. The hollow glass tube contains a calibrated scale showing the specific gravity, degrees balling (percent sugar) or proof, depending on the type of hydrometer. More accurate readings can be obtained by using two specialized hydrometers rather than a single all-purpose instrument.

the color, the greater the amount of unconverted starch remaining in the mash.

If the malting is incomplete, continue stirring for another 10 to 15 minutes and then test again. If it's still incomplete, add 10 percent more malt or enzyme and agitate for an additional 15 to 20 minutes.

Pre-malting

A technique known as pre-malting can be used to increase the efficiency of the conversion process. Pre-malting means adding a small amount of malt or enzyme, about 10 percent of the total to be used, to the mash *prior* to cooking. This starts the starch-to-sugar conversion in motion so that when the rest of the malt is added the conversion will be much more complete.

Pre-malting also serves another very important function. It prevents thickening that sometimes occurs when the mash is cooked. The starch in the mash might cause the soup to take on the consistency of gelatin. This makes it difficult to stir and handle. Pre-malting liquifies enough starch to prevent the thickening. Therefore, it is highly recommended for crops such as corn.

MAKING MASH FROM A CELLULOSE FEEDSTOCK

Although the basic process for converting cellulose to glucose is similar to malting, the process is hindered by the molecular configuration of cellulose. The long chains of simple sugars that make up the typical cellulose molecule twist and turn in such a way that they form internal bonds that are highly resistant to chemical attack. This is what makes cellulose a tough substance to decompose. It can be done, but for the effort involved sugar and starch feedstocks return a much better yield of alcohol for the do-it-yourselfer.

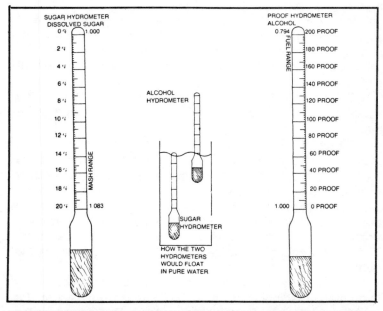

Fig. 4-2. By taking a single all-purpose hydrometer and making it into two specialized hydrometers, more accurate readings are possible. The sugar hydrometer shown above is for example only. Most come with scales that read from 1.000 to 2.000 or 0 to 50 degrees Balling. Proof hydrometers might also show the alcohol content in percent or a combination of proof and percent.

On a commercial scale, the cellulose feedstock is shredded and mixed with water to form a thin slurry (10 percent solids and 90 percent water). Acid and heat, or heat and enzymes, are then used to hydrolyze the cellulose molecules into glucose. The tearing apart of the cellulose chains frees the sugars for fermentation.

CHECKING SUGAR CONTENT

Because the sugar concentration plays such a critical role in fermentation, a means of measuring the amount of dissolved sugar in the mash is absolutely essential. The best way to accomplish this is with a hydrometer (Fig. 4-1 and 4-2).

A hydrometer is a simple instrument that measures the specific gravity or density of a liquid. It consists of a small glass float with a weight in one and a scale running the length of the oblong float. To test a liquid, drop in the float and read the number off the scale where the float floats. It's no more difficult than reading the waterline on a ship.

You can use a hydrometer to determine the amount of sugar in the mash because dissolved sugar makes the water heavier (more dense). The specific gravity, therefore, increases with the amount of dissolved sugar. This causes the float to float higher.

The important point to remember is that the higher the specific gravity, the greater the amount of dissolved sugar in the mash.

To test the mash, draw off a small sample of liquid and allow it to cool to room temperature. Immerse the hydrometer in the liquid and let the float seek its own level. The specific gravity can be read off the scale at the point where the bulb floats.

For example, if you test a small sample of mash and find that the hydrometer reads 1.0568, look up 1.0568 in Table 4-2 and this tells you that you've got a 14 percent sugar solution. That is close enough to the ideal so that you can add the yeast to begin fermentation.

There are a few items that should be mentioned about hydrometer readings. First of all, the figures in Table 4-2 are for liquids at 68 degrees F (room temperature). This is because the specific gravity of water is defined as 1.000 at 68 degrees F. All the other readings are based on this figure. As the temperature increases, water becomes less dense. This causes a decrease in the hydrometer reading from what it should be. Conversely,

Table 4-2. Specific Gravity and Sugar Content.

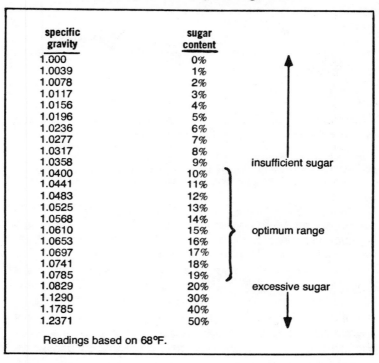

specific gravity	sugar content	
1.000	0%	
1.0039	1%	
1.0078	2%	
1.0117	3%	
1.0156	4%	
1.0196	5%	
1.0236	6%	
1.0277	7%	
1.0317	8%	
1.0358	9%	insufficient sugar
1.0400	10%	
1.0441	11%	
1.0483	12%	
1.0525	13%	
1.0568	14%	
1.0610	15%	optimum range
1.0653	16%	
1.0697	17%	
1.0741	18%	
1.0785	19%	
1.0829	20%	excessive sugar
1.1290	30%	
1.1785	40%	
1.2371	50%	

Readings based on 68°F.

as water becomes cooler, it becomes more dense. This increases the reading over what it should be. What all this means is that hydrometer readings must be "corrected" for temperatures other than those given in Table 4-2. Fortunately, the complicated mathematical equations can be avoided since most hydrometers come with built-in thermometers. The reading can then be corrected according to the scale on the thermometer.

If all this information on specific gravity and correction factors has you fumbling for your pocket calculator, fear not. There are hydrometers available that eliminate the need for confusing numbers and conversion tables. These are the hydrometers that come calibrated in percentage of sugar, called *degrees Balling*. One degree on this scale equals 1 percent dissolved solids. For example, if the hydrometer reads 14 degrees Balling, then the mash would have a 14 percent sugar concentration.

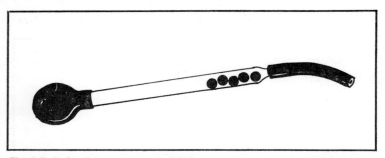

Fig. 4-3. A simple mash tester made from an automotive antifreeze tester. If one ball floats, the sugar content is 10 percent. If two balls float, the sugar content is 15 percent. If three balls float, the sugar content is 20 percent. The mash should have enough sugar to cause one or two balls to float. If three or more balls float, the mash has too much sugar.

Balling hydrometers usually come with a range of 1.000 to 2.000, making them more than adequate to handle any mash. This type of hydrometer is also referred to as a "Brix" hydrometer. Adolf F. Brix, a German scientist, developed it.

Hydrometers can be purchased from laboratory equipment supply houses or from the suppliers listed in the Appendix. Prices range from around $5 for a Brix type hydrometer up to $25 or more for a laboratory model. You will also need a "proof hydrometer" to measure the percentage of alcohol in your finished product. Although you might be able to use some hydrometers to measure both sugar concentration and proof, most don't have that kind of range. This is because sugar solutions are heavier than water and alcohol solutions are lighter than water. Using a different hydrometer for each provides more accurate results since it can be calibrated to be more sensitive within the limited range being measured. Proof hydrometers sell for approximately the same price as sugar hydrometers.

AN INEXPENSIVE HYDROMETER

If you are only experimenting and aren't ready to buy a "real" hydrometer just yet, you can use an ordinary antifreeze tester (Fig. 4-3) to get a rough indication of the amount of antifreeze protection in the radiator. Since both ethylene glycol antifreeze and dissolved sugar increase the specific gravity of water, the number of floating balls can be used to judge the amount of sugar in the mash.

Based on some backyard experiments, I found that a 10 percent sugar solution caused one ball to float. This corresponded with antifreeze protection to +20 degrees F. A 15 percent sugar content caused two balls to float. This equalled antifreeze protection to +5 degrees F. A 20 percent sugar concentration made three balls float. This equalled –10 degrees F on the tester. Therefore, if the mash is checked with an antifreeze tester such as this, an optimum sugar level of 15 percent would correspond to a +5 degree F reading on the tester. Perhaps someone will come out with such a tester already calibrated to read sugar content. It would certainly make things a lot easier for the do-it-yourselfer.

MASH PREPARATION

Keep in mind that all recipes are approximations, and that the exact time for malting or the exact amount of water required to dilute the mash to the proper consistency can vary somewhat from situation to situation. One do-it-yourselfer's batch of corn can vary from another's in moisture content, starch content and sugar content. These factors depend on the strain or variety of crop planted, the particular growing conditions and soil quality of the area, the moisture content at time of harvest and so on. For the best results, start with the recommended proportions and fine tune from there on trial runs.

Fortunately, the variations in crops are usually not enough to make a significant difference in the recipes. During preparation, the mash can tolerate a fair amount of variation in the relative proportions of water, crop and malt. Mistakes usually can be corrected. A mash that is too thick can be thinned by adding more water. A mash that is too thin can be corrected by evaporating excess water. Incomplete starch conversion can be improved by allowing more time or using more malt or enzymes. The only mistakes that can ruin the mash are those involving sugar concentration, pH and malting temperature. Too high a temperature during malting (over 170 degrees F for barley malt) will destroy the sugar making ability of the enzymes. This means that the malting process would have to be repeated. Dumping in too much acid to lower the pH might also push the mash beyond the limits of rebound. Remember that a little acid goes a long way. Too high a sugar concentration (over 20 percent) in the finished mash might also foul up fermentation. Less than 10 percent sugar might make fermentation unproductive.

As long as you end up with a mash containing about 15 percent simple sugar and a pH between 4.5 and 5.0, the mash should be suitable for fermentation. Once the mash is ready, proceed on to fermentation as quickly as possible to avoid contamination by unwanted organisms.

Fruits

Press the juice from the crop.

Check for sugar content of at least 10 percent. If it is too diluted, evaporate some excess water.

Check the pH. Add acid or backslop with stillage up to 25 percent to increase acidity to a level of 4.5 to 5.0.

Add yeast at 2 percent per 1000 gallons of juice to ferment.

Sugar Cane

Mulch the sugar cane.

Add about 2 parts water for 1 part sugar cane.

Heat to not quite a boil to dissolve sugar in the cane.

Drain off the water that contains sugar and repeat the process until most of the sugar is extracted.

Check the sugar content of the washing water. If it is too concentrated or too diluted, add or evaporate water to achieve the desired 15 percent concentration.

Check the pH. Add acid or backslop with stillage up to 25 percent to increase acidity to a level of 4.5 to 5.0.

Add yeast at 1½ to 2 pounds per 1000 gallons of liquid to ferment.

Sugar Beets

Press juice from the crop or crush the beets and add 1 percent to 2 percent malt by weight to convert starchy pulp to sugar. Heat to 145 degrees F for 30 minutes if you are using malted pulp. Add enough water to form a thick soup.

Check the sugar content (between 10 percent and 18 percent). Dilute if necessary with additional water.

Check the pH. Add acid or backslop with stillage up to 25 percent to increase acidity to a level of 4.5 to 5.0.

Add yeast at 2 pounds per 1000 gallons of juice to ferment.

Molasses

Check the sugar content of the molasses and add water to dilute the mash to the desired 15 percent range or backslop with stillage up to 50 percent and then add water to achieve the desired dilution.

Check the pH. Add acid (or backslop) to increase acidity to a level of 4.5 to 5.0.

Add yeast at 2 pounds per 1000 gallons of mash to ferment.

A note on sugar mashes. Backslopping with stillage residue serves two purposes. It increases the acidity of the mash, which sugar feedstocks usually require, and it adds nutrients to the mash that might be lacking. These nutrients will aid yeast growth during fermentation.

Corn and Grains

Corn and other grains must be milled, mixed with water, pre-malted, cooked and malted prior to fermentation. See the checklist for mash preparation for the exact steps.

Most grain mashes will have a low enough pH not to require additional adjustment with acid or backslopping. The mash should also contain plenty of sugar and nutrients for healthy yeast growth. Backslopping is unnecessary.

Potatoes

Since potatoes are about 80 percent water, mashing and the addition of a slight amount of water are all that is necessary to prepare the soup for cooking.

Pre-malt with 10 percent of the barley malt or enzymes and stir well.

Cook for two to three hours.

Cool the mash to 145 degrees F and add only enough water to make up for that which is lost during cooking.

Add the remaining malt—3 to 4 pounds of malt per 100 pounds of potatoes—and agitate for 15 to 20 minutes.

Test with iodine for completeness of the starch conversion. If incomplete, add more malt or allow more time.

Check the sugar content and pH for the desired range. Cool and ferment as usual.

Sweet Potatoes

The procedure is the same as with regular potatoes only more water is needed to dilute the potatoes to a thick soup-like consistency. Sweet potatoes contain about 66 percent moisture versus 80 percent for regular potatoes.

Pre-malt, cook and malt as before.

Test for thoroughness of starch conversion.

Test sugar concentration and pH. The sugar should be about 15 percent and the pH should be 4.5 to 5.0.

Add yeast to ferment sugars into alcohol.

Other Starch Feedstocks

The basic procedure is to grind or mash the crop and add enough water to form a thick soup.

To pre-malt, cook and malt as before. The amount of malt used is determined by trial runs and iodine tests. The recommended starting dosage for corn is 10 percent of the weight of the crop for barley malt and 1 percent of the weight of the crop for commercial enzymes.

Test the sugar concentration and dilute if necessary to the desired 15 percent range.

Test the pH and adjust with acid to a level of 4.5 to 5.0.

Add yeast to ferment sugars into alcohol.

Checklist for Mash Preparation of a Starch Feedstock

- Grind for starch feedstock to a consistency of sand.
- Mix with water in a ratio of 2 parts water to 1 part solids. Example: 1 bushel equals 8 gallons. Add 16 gallons of water to each bushel of ground grain.
- Check pH if you are using commercial enzymes to pre-malt mash. It should be within the range recommended by the supplier.
- Add 10 percent of the enzymes or malt to pre-malt the Mix well.
- Cook the mash at a slow boil for 30 minutes.
- Cool and dilute the mash by adding more water in a ratio of 2 parts water for each 1 part solids in the mash. Example: Add another 16 gallons of water for each bushel of grain already in the mash.
- Mash temperature should be approximately 145 degrees F.
- Add the remaining portion of barley malt or enzymes. Malt should be added in the ratio of 10 percent to 15 percent of the weight of the solids in the mash. Enzymes should be used in the ratio of 1 percent of the weight of the grain. Example: Corn weighs about 56 pounds per bushel. If you are using barley malt, use about 10 percent to 15 percent of that figure per bushel, or 5.6 to 8.4 pounds of malt. With enzymes, it would be 1 percent or about one-half pound per bushel.
- Agitate the mash for 30 minutes. Starch to sugar conversion should now be complete.
- Test for the presence of unconverted starch by using an iodine solution. If the iodine reacts with a small sample of the mash, starch is still present. Continue agitation for an additional 10 to 15 minutes or add 10 percent additional malt or enzymes.
- Once all the starch has been converted to fermentable sugar, test the sugar concentration with a hydrometer. It should be within the 15 percent range of optimum yeast growth. If it is too high, dilute with water.

- Test the pH of the mash. It should be between 4.5 to 5.0. Add diluted acid if the pH is too high.
- Allow the mash to cool to about 80 degrees F or 85 degrees F before adding yeast.

Municipal Garbage

Shred the trash and remove non-cellulose materials such as metals, glass, rocks and plastics. On an average, garbage is about 35 percent to 55 percent paper.

Fine shred the remaining cellulose materials (mostly paper).

Mix with water to form a slurry. Use 9 parts water to 1 part shredded solids.

Add acid to form an 8 percent to 10 percent acid solution and cook under pressure at 360 degrees F to 440 degrees F to break cellulose down into fermentable sugar.

Or add Cellulase enzymes and heat to 150 degrees F to break down cellulose into fermentable sugar.

Cool the slurry to 80 degrees F.

Check the sugar content of the slurry and remove excess water as needed to achieve the desired 15 percent sugar concentration.

Check the pH and adjust accordingly for a level of 4.5 to 5.0.

Add yeast and ferment the sugar into alcohol.

Other Cellulose Feedstocks

Shred the feedstock to resemble confetti.

Mix with water to form a dilute slurry as described for municipal garbage and follow the same procedures.

Chapter 5
Fermentation

Once a simple sugar solution has been prepared from the basic feedstock, the mash pH balanced for optimum yeast growth (4.5 to 5.0) and cooled to 75 degrees F to 85 degrees F, the mash is ready to be fermented into ethanol alcohol. For this you will need the assistance of some very special microorganisms known as yeast.

Yeast are a type of unicellular fungi that feed on dissolved sugar to produce ethanol alcohol and carbon dioxide gas as byproducts of digestion. Yeast are able to convert sugar into ethanol because they secrete enzymes into the mash. The enzymes help the yeast digest sugar and also cause a chemical reaction that transforms simple sugar into alcohol and CO_2. The particular enzyme that is responsible for this amazing feat is *zymase*.

$$C_6H_{12}O_6 + \text{zymase} = 2\ C_2H_5OH + 2\ CO_2$$
$$\text{(glucose)} + \left(\begin{array}{c}\text{yeast}\\\text{enzyme}\end{array}\right) = \left(\begin{array}{c}\text{ethanol}\\\text{alcohol}\end{array}\right) + \left(\begin{array}{c}\text{carbon}\\\text{dioxide}\end{array}\right)$$

The zymase enzyme secreted by the yeast causes each sugar molecule to break down into two ethanol alcohol molecules and two molecules of carbon dioxide. The alcohol remains in the mash and gaseous CO_2 bubbles out.

Once the mash begins to "work," it will bubble and foam as yeast multiply and feed on the sugar. The rapid increase in chemical activity also causes the temperature to rise. As long as the mash is between 72 degrees F and 90 degrees F, fermentation should proceed without problems. The higher the temperature, the faster the yeast will work—up to a point. Be careful not to allow the mash to rise above 95 degrees F during fermentation otherwise the yeast might be destroyed. Yeast are very sensitive to temperature changes. Going much higher than about 90 degrees F runs the risk of killing your little fuel producers.

The best temperature to aim for is approximately 85 degrees F. This leaves a 10 degree safety margin yet it is high enough to ensure vigorous growth and maximum fuel production in a reasonable amount of time. Cooler than this is satisfactory but the lower the temperature, the slower the fermentation process. Around 40 degrees F, the yeast stop working altogether—until the temperature is increased to start the mash working again.

FERMENTER COOLING

Overheating is more of a problem in large fermentation vessels where the surface-to-volume ratio limits the amount of heat that can escape from the mash. If the size of the fermenter is limited to that of a 55 gallon drum or trash can, overheating should not be a problem unless the weather is unusually hot. In that case, it might be best to wait until the weather cools before beginning fermentation or you could place the fermenter in a cool location such as a basement or root cellar. Be sure to provide some ventilation if the fermenter is placed in a small room or building since excess CO_2 will tend to accumulate. Because carbon dioxide is heavier than air, it will sink to the floor as it comes out of the fermenter. This creates no problems if the fermenter is placed outside or in an above ground building— provided the building has an open door, window or some type of venting. In an underground room without adequate ventilation, CO_2 might accumulate until it excludes the oxygen you need to breathe. Carbon dioxide itself is non-poisonous and non-explosive, but breathing it exclusively can result in suffocation.

If the fermenter should become too warm, the easiest way to cool it is to hose off the outside with cool water. A more sophisticated (and expensive) approach is to wrap cooling coils either around the outside of the tank or inside the fermenter. When the temperature exceeds 88 degrees F to 90 degrees F, a thermostat opens and cool water circulates through the plumbing to cool the mash. This is more practical than sitting around and watching a thermometer.

External cooling coils can be made by wrapping an old garden hose, plastic or metal tubing around the tank. Any coils that go inside the tank should be an acid resistant plastic non-corrosive or coated metal, such as stainless steel, copper or epoxy painted steel. Internal coils should also be constructed so that they are easy to clean.

Another technique sometimes employed by do-it-yourselfers is ice cooling. So as not to dilute the mash with additional water, the ice is placed in a sealed plastic bag and then lowered into the mash. After the mash has cooled to a safe temperature, the bag is removed. Although simple, this technique risks the danger of contamination. Unless the plastic bag is absolutely clean, unwanted organisms might hitch a ride into the fermenter and spoil the mash.

SEALED FERMENTATION

Once fermentation is underway, the fermenter should not be opened. Opening not only risks contamination but it also permits fresh oxygen to enter the tank. When an excess of oxygen is present, enzymes in the mash might convert the sugars into vinegar instead of alcohol. Internal combustion engines won't run on acetic acid (vinegar).

To keep unwanted organisms and excess oxygen out of the fermenter, the tank must be sealed to the outside atmosphere—except for a small vent through which the waste CO_2 can escape. By adding a simple water trap to this vent, it becomes a *fermentation lock*.

The function of the fermentation lock (Fig. 5-1) is to seal the mash from the outside environment while allowing the gas within the fermenter to escape. As gas pressure builds up inside, it begins to bubble out through the water trap. The slight resistance offered by the water keeps just enough pressure within the fermenter to maintain a one-way flow through the lock. Unwanted organisms and oxygen can't get in because the CO_2 is trying to get out.

The fermentation lock itself can be of any design so long as it has a U-shaped bend filled with water. It should resemble your kitchen sink. Just be sure the tubing is large enough so that pressure doesn't build up and blow the lid off the fermenter. This has been known to happen.

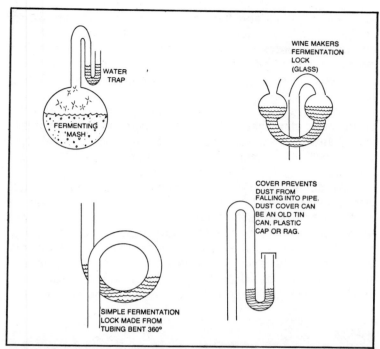

WATER TRAP

FERMENTING MASH

WINE MAKERS FERMENTATION LOCK (GLASS)

COVER PREVENTS DUST FROM FALLING INTO PIPE. DUST COVER CAN BE AN OLD TIN CAN, PLASTIC CAP OR RAG.

SIMPLE FERMENTATION LOCK MADE FROM TUBING BENT 360°

Fig. 5-1. Fermentation locks allow CO_2 to escape from fermenting mash but prevent airborne organisms from contaminating the mash.

FERMENTER CONSTRUCTION

The fermenter vessel should be made from a material that is both resistant to acid and easy to clean. Commercial plants use stainless steel or copper (Fig. 5-2). While these materials are probably too expensive for most backyard experimenters, a reasonably good fermenter can be made by coating the inside of a 55 gallon drum (Fig. 5-3) with epoxy paint. Fiberglass or hard enamel will also work as a protective coating. Don't use a lead based paint. The acids will leach the lead into the mash. A little lead in the fuel won't make a difference, but it could poison livestock if there were much of it in the mash residue. Many a moonshiner has experienced the ill effects of lead poisoning from crude home-built stills that had lead soldered joints or seams.

An easier and safer approach is to use a large plastic trash bag to line the barrel. The plastic eliminates the need to coat the inside of the barrel—(as long as the bag doesn't spring a leak)—and it eliminates the need to clean the barrel between batches. The bag should be disposed of after each batch and not reused in order to avoid any danger of contamination. If a plastic liner isn't used, the fermenter barrel or tank should be scrubbed after each batch. This is especially important during humid summer weather.

Another container often used as a fermenter is a large plastic trash can. They are relatively inexpensive, readily available, lightweight and can be stacked for storage. It can be used with or without a plastic liner. If no liner is used, it must be cleaned between batches.

Whatever type of fermenter is chosen, it should be fitted with a tight sealing lid and fermentation lock. The lid can be sealed with flexible calk, or weatherstripping if it is necessary to make it air tight. Remember, you want to keep oxygen and other airborne items out of the mash. If a plastic liner or container is used, a separate vessel will be needed for cooking. A stainless steel, copper or coated steel kettle or drum can be used. Otherwise, the mash can be cooked, malted and fermented all in the

Fig. 5-2. Fermentation tanks at the Archer Daniels Midland alcohol plant in Decatur, Illinois.

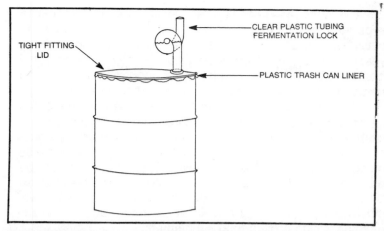

Fig. 5-3. A simple fermenter made from a 55 gallon drum. The mash is first cooked and malted in a cooking kettle and then poured into the fermenter. Half a pound of yeast is mixed into the mash before the lid is put in place. The fermenter will then work for about three days. When gas ceases to bubble out through the vent, the beer is ready to be separated from the mash for distillation.

same tank. As fermentation continues, the mash will bubble vigorously as CO_2 is being released. At the same time, the concentration of alcohol will be gradually increasing. Eventually an increasingly toxic concentration of alcohol and a diminishing supply of sugar will begin to take its toll on the activity of the yeast. After about three days—the time varys depending on the feedstock, type of yeast, growing conditions—the mash will simmer down and cease to bubble. The yeast cake that formed on top will sink to the bottom of the fermenter. When this happens, fermentation has ended.

Because the mash has finished working doesn't mean that 100 percent of the available sugar has been consumed by the yeast. What usually happens is that the yeast succumb to the toxic effects of their own by-products of digestion (such as alcohol) before they can "starve to death" from a lack of sugar. In other words, they pollute themselves to death. This happens to baker's yeast when the alcohol concentration reaches 8 percent to 12 percent, or 10 percent to 14 percent or more for brewer's yeast. It all depends on the strain of yeast and its resistance to alcohol. Because of this, brewer's yeast is preferred over baker's yeast for fuel production. Brewer's yeast will give a higher alcohol yield.

When the mash has finished working, the beer should be separated from the mash as soon as possible. If it exposed to outside air and allowed to sit for any length of time, the surviving yeast and enzymes might begin to convert some of the alcohol into vinegar. Because of this, it's a good idea to keep the fermenter sealed until you're ready to strain the mash.

To separate the beer from the mash, the contents of the fermenter should be strained through a fine screen or porous cloth such as burlap. Old time moonshiners preferred an ordinary bedsheet. Regardless of the filter used, try to separate as much liquid from the mash as possible. The beer can then be distilled to recover the alcohol. The leftover mash residue can be used for backslopping the next batch, for fertilizer, for livestock feed in a 50—50 ration with their normal feed, or dried and used as a high protein food supplement.

YEAST

Baker's yeast and brewer's yeast are the two main varieties of yeast. Baker's yeast is used for baking and brewer's yeast is used for brewing alcohol. Baker's yeast is valued for its ability to produce carbon dioxide gas. The bubbles are what causes bread to rise when yeast is added to the dough. When the bread is baked, the heat evaporates the alcohol and hardens the dough.

With alcohol production, brewer's yeast is the choice because of its higher tolerance (and higher yield) for alcohol. This results in more alcohol from a given batch of mash. Although either type of yeast will work for the do-it-yourselfer, brewer's yeast should be your first choice.

Sources

Baker's yeast is readily available in grocery stores, bakeries and from bakery supply houses in dry or moist cake form. Brewer's yeast, unfortunately, is still rather difficult for the do-it-yourselfer to obtain. This situation could change for the better if enough people get into home fuel production. There has not been enough of a demand outside the brewing industry to create a strong market for brewer's yeast. Commercial operations currently either grow their own yeast or buy it in large quantities from suppliers such as Anheuser-Busch, Universal Foods or Standard Brands. Prices for 50 pound quantities range from about 30 cents a pound to 80 cents a pound.

Although any of the big suppliers would probably sell you a 50 pound sack of brewer's yeast (also called distiller's yeast), there might be a problem in getting it delivered. Yeast must be kept under refrigeration to prevent spoilage. The closer to 32 degrees F the better, but not below freezing. This means it has to be shipped in a refrigerated truck and stored in a refrigerated compartment until you can take delivery. It should then be placed in a picnic cooler while being transported home. Store the yeast in your refrigerator until it's needed. Shelf life will depend on the age of the yeast when purchased, the care it's received in handling and packaging. Most commercially bought yeast will have an expiration date stamped on it. The best advice, however, is don't buy the yeast too far ahead of time.

A local distillery is another potential source of brewer's yeast. However, they might or might not be willing to sell you a small quantity. If they're willing to part with some at a price you can afford, use the same precautions in handling and storage to prevent spoilage.

Another source might be a local wine shop that sells home wine making supplies. Although they might carry a stock of brewer's yeast, they will probably have to special order it in the quantities you will need for fuel production.

If you find brewer's yeast is unavailable in your area or that it's just too much of a problem to have it shipped in from a major supplier, you can always use baker's yeast.

Yeast Dosage

One pound of yeast should be sufficient to ferment 500 gallons of mash. When estimating your yeast dosage, figure about 0.2 percent to 1 percent of the weight of the original dry weight of a starch feedstock. For example, to each bushel of corn in the mash you would add 0.2 percent to 1 percent of the weight of that bushel (which is about 56 pounds), or about one-tenth to one-half pound of yeast.

For best results, both dry yeast and cake yeast should be dissolved in a small quantity of water at about 90 degrees F and allowed to sit for about 15 minutes before it's added to the mash. This "reactivates" the yeast. If you use a gallon jug for this purpose, you might also add a few tablespoons of sugar to help get the yeast going—but don't overdo it. Remember, too high a sugar concentration can be harmful to the yeast.

After pre-mixing the yeast with water, blend it in thoroughly with the mash to evenly distribute the yeast cells. A few minutes of stirring should do the trick. Then sit back and let the yeast do its thing.

Although pre-mixing the yeast with water is recommended, it's not absolutely essential. In fact, you can add the dry yeast to the fermenter just before you fill it with mash. Just remember to stir it before sealing the lid.

BATCH FERMENTATION

Two ways to speed fuel production is to use "continuous" fermentation equipment or to use several fermenting tanks instead of just one. Continuous fermentation, unfortunately, is pretty much limited to the large scale commercial plants because of the complexity and costs. Mash is cooked and malted on a continuous basis and pumped into a number of large fermenters. After each fermenter has completed its job, the beer is pumped out, one tank after the next, and distilled to make alcohol on a more or less continuous basis.

This same concept can be used by the do-it-yourselfer to increase production. The advantage of using multiple fermenters is that several batches of mash can be prepared in rapid succession. As each batch is readied, it is poured into a fermenter. In other words, rather than preparing and fermenting only a single batch of mash, you can prepare and ferment several batches at one time.

A typical setup for a small fuel producer might include a single 55 gallon drum for a cooking vessel and six similar drums for fermenters. As soon as a batch of mash has been cooked and malted, it is dumped into a fermenter, charged with yeast and sealed. Then another batch is prepared and so on until all the fermenters are full. There will be the equivalent of six runs worth of beer at the end of the fermentation period. The beer can then be distilled as it is needed.

You can get the same results by using a larger cooking vessel and fermentation tank to increase production. But if the larger tanks prove to be too expensive for your needs, the smaller drums or trash cans will work just as well. The use of multiple fermenters also reduces the odds of losing the entire run through contamination. Something might get into one of the tanks to foul things up, but it's unlikely you'd lose them all. Multiple fermentation also gives you the flexibility to experiment within a single run. For example, you might try adding various amounts of yeast to the different tanks to see which dosage yields the best results. The same goes for trying different strains of yeast, playing with the temperature, pH or sugar content. After you have gained experience and narrowed down the combinations that work best, you can think about moving up to larger tanks and increasing your fuel production.

Precautions

Keep in mind that the purpose of your alcohol production equipment and experimental permit is to make fuel, not booze. A lot of things can happen during fermentation, including the production of small amounts of butanol, propanol, aldehydes and ketones. These by-products have no effect on the use of alcohol as a motor fuel, but they can be deadly if consumed. Beverage grade alcohol must be made under the strictest of sanitary conditions and then carefully distilled to remove all traces of fusel oil and other potentially harmful chemicals. If the U.S Bureau of Alcohol, Tobacco & Firearms finds out that you've been moonshining instead of making fuel, you will have a serious problem.

Bacterial Fermentation

According to Dr. Ljungdahl of the University of Georgia, a new microorganism discovered living in the hot springs of Yellowstone National Park might prove to be more efficient than yeast for converting sugar into ethanol. The new bacteria has a real tongue twister of a name, *thermoanaerobactor ethanolicus*.

The bacteria has the unusual ability to grow at temperatures as high as 172 degrees F. Therefore, it could ferment sugars on a continuous basis in a high temperature commercial operation. This would enable the do-it-yourselfer to ferment the mash right after cooking without worrying about cooling the mash to the proper temperature. It would also eliminate the need for cooling coils in large fermentation tanks. Another benefit is that the higher temperatures reduce the possibility of contamination by other organisms because few can survive in such a hot environment. Dr. Ljungdahl also reports that the new bacteria is more tolerant of a wide range in pH, making it easier to work with than yeast.

Table 5-1. Comparison of Feed Values.

	corn	DDG	soybean meal
Dry matter (%)	88.0	93.8	89.6
Crude protein (%)	8.9	27.0	44.0
Crude fat (%)	3.5	9.0	0.5
Crude fiber (%)	2.9	13.0	7.0
Ruminant digestible protein (%)	5.8	19.3	37.5

The greatest potential for the new bacteria, however, is for converting cellulose into alcohol. When combined with another type of bacterium (*Clostridium thermocellum*) that attacks celluslose, the new bacteria accelerates the rate of breakdown while producing alcohol. This could give the large-scale conversion of municipal garbage, scrap paper and biomass residue the boost needed to become competitive with starch and sugar feedstocks. It might also enable the small-scale producer to economically manufacture fuel from such resources.

STILLAGE FEED VALUE

The feed value of the residue from grain alcohol fermentation is an important factor in the overall economics of do-it-yourself fuel production. This by-product is called distiller's dried grains (DDG) and it is a well established product that is marketed by commercial distilleries. The product contains basically the same vitamins, proteins and minerals as the original feedstock and is sold as a high protein food supplement. Its main competition is soybean meal.

After fermentation, each bushel of corn leaves about 17 to 18 pounds of high protein stillage (dry basis). The U.S Department of Agriculture used the information in Table 5-1 to compare the feed values of whole corn, DDG from corn and soybean meal.

Distiller's grain from an on-farm alcohol plant has advantages over commercial distilleries. The big plants spend considerable amounts of energy to dry the protein byproduct. A farmer can feed the supplement wet—storage time is up to a week. In addition two sets of transportation costs are eliminated—moving the corn to the distillery and then moving the by-products back to the farm.

Chapter 6
Distillation

The basic process of distillation is to boil off the alcohol and recondense it, leaving most of the water behind. Any still you can conceive will work so long as it will:

—Boil the beer.
—Separate the alcohol and water vapors.
—Condense the vapors into liquid.

The type of still design or energy source to power it is limited only by your imagination. One enterprising do-it-yourselfer came up with a "microwave" powered still. Electricity from a windmill generator powered a microwave heating element which in turn boiled the liquid inside his "microwave still."

SEPARATION

Once fermentation is complete and the beer has been strained from the mash residue, you are ready to separate the alcohol. This is the process known as distillation (Fig. 6-1) and it accounts for roughly 50 percent to 80 percent of the heat energy that goes into making alcohol.

If water and alcohol were as dissimilar as water and oil, separating the alcohol would be simple. You'd just let the beer settle and then skim off the alcohol layer. The trouble is that although water and alcohol have different densities (alcohol is lighter than water), they mix quite readily. Once mixed together, they prefer to stay that way. Recovering the alcohol requires the application of some basic physics.

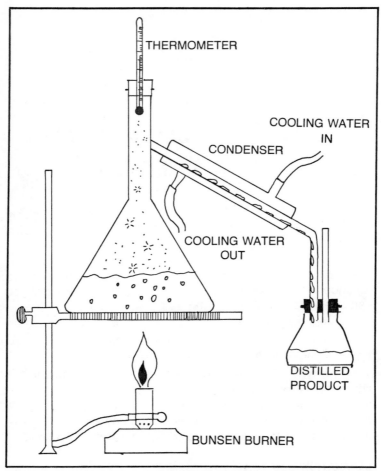

THERMOMETER

COOLING WATER IN

CONDENSER

COOLING WATER OUT

DISTILLED PRODUCT

BUNSEN BURNER

Fig. 6-1. The alcohol containing liquid is brought to a slow boil, driving off a large portion of the alcohol as vapor. The hot vapors rise and enter the condenser where they are cooled and change into liquid again. Starting with a 8% alcohol solution in the distilling flask, the finished product would be about 40% to 45% alcohol. Running the distilled product through the apparatus again would increase the content to 65% to 70%. Repeating this process over and over would gradually increase the alcohol to a maximum concentration of about 95%.

The property that will enable you to separate the alcohol from the water is simply the lower boiling point of alcohol. Water boils at 212 degrees F. Pure ethanol alcohol, on the other hand, boils at 173 degrees F—a difference of 39 degrees. That might not sound like much, but it is enough to make separation through distillation possible. By gradually heating the beer to a slow boil, you will drive off more alcohol vapor (Fig. 6-2) than steam because alcohol has a lower boiling temperature. This vapor can then be collected, cooled and condensed into a liquid. By repeating this process over and over, you will eventually obtain nearly pure (95 percent) alcohol.

The beer will begin to boil at a temperature somewhere between 173 degrees F and 212 degrees F, depending on the relative amounts of alcohol and water in the liquid. The higher the percentage of alcohol, the closer the boiling point will be to 173 degrees F. Conversely, the higher the percentage of water, the closer to 212 degrees F the boiling point will be. For example, an 8 percent alcohol solution will boil at about 200 degrees F because of the large proportion of water present.

As the beer is boiling, the amount of alcohol remaining in the liquid will gradually diminish as the vapors boil off. This causes the boiling temperature of the liquid to steadily rise. When only water remains, the liquid will be boiling at 212 degrees F and the vapors will be pure steam. In other words, the vapors which boil off first will contain a much higher proportion of alcohol than those which boil off last. The alcohol rich vapors

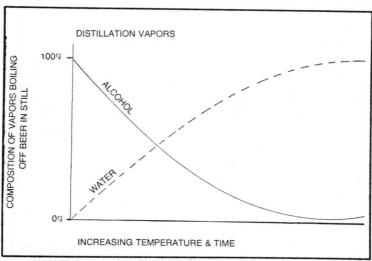

Fig. 6-2. As the beer begins to boil, the vapors will be mostly alcohol. But as time passes, the percentage of water will steadily increase—also causing the vapor temperature to gradually rise. When 200 degrees F is reached, most of the alcohol will have been recovered. The remaining vapors will be mostly water.

that are generated at the beginning are sometimes called the *heads*. Those that contain much less alcohol towards the end of the run are the *tails*.

You might think that by carefully heating the beer to 173 degrees F, you could boil off only the alcohol and not the water. However, you can't fool Mother Nature. When water ánd alcohol are mixed together, the molecules interact and cling to one another. If such a mixture is heated to 173 degrees F, vary few alcohol molecules will be able to break free. The only way to break up this relationship is to bring the liquid to a boil. This will free much of the alcohol and also a certain amount of steam— even though the temperature is less than 212 degrees F.

When an 8 percent solution begins to boil at 200 degrees F in a still, the vapors coming off will average between 175 degrees F and 185 degrees F. The difference in temperature between the liquid and vapor occurs because it takes a certain amount of energy to make the transition from a liquid to a gas. As the alcohol and water vapors rise, some of the water will begin to condense into droplets and fall back into the liquid because it isn't hot enough to keep on going. Nevertheless, a large amount of water vapor will still be generated because of the large proportion of water in the liquid.

The temperature of the vapors will gradually increase as the amount of alcohol being boiled off decreases. When the vapor temperature reaches about 200 degrees F, the liquid will be boiling at close to 212 degrees F—signalling that most of the alcohol has been boiled off. At this point, the vapor coming off is mostly steam. The still should be shut down to avoid overdilluting the distilled alcohol with excessive water. Remember, as the vapor temperature rises, the amount of water in the vapor also rises. Since you don't want the tails to dillute the fuel, you either shut down the still or collect the tails separately for redistillation. By running this diluted fuel through the still again, the excess water will be removed.

Methods of Distillation

The two basic approaches to distillation are batch processing and continuous processing. Simple batch processing involves pouring a quantity of beer into a still and boiling it until most of the alcohol has been removed. Continuous distillation involves boiling a continuous stream of beer to recover the alcohol as it passes through. By varying the rate of flow, the percentage of alcohol recovered can be regulated. This is the process used by most commercial producers. Because of the difficulty in controlling distillation on a continuous basis, most do-it-yourselfers would be well advised to stick with the tried-and-proven batch technique.

Batch distillation can be subdivided into simple and complex distillation. Simple distillation is boiling the beer to drive off the alcohol vapors and then recondensing these vapors into a liquid. It's simple enough to do but it doesn't yield as high a proof alcohol as complex distillation. This is because a lot of steam boils off with the alcohol, and if no other means is employed to remove the excess water, the end product will not contain more than about 40 percent alcohol (or about 80 proof). This is potent enough for liquor, but not for motor fuel. For that, you will need at least 80 percent or 160 proof.

To increase the proof, the fuel from simple distillation can be run through the still a second time. Redistillation would result in an increase in the alcohol content to about 65 percent to 70 percent or 130 to 140 proof. A third pass would raise the concentration high enough to be worthy of being called a motor fuel. Repeated passes through the still wastes a lot of energy and effort. It makes more sense to add a device called a *column* to the still to increase the proof the first time through. This is known as complex or *reflux* distillation.

Distilling Columns

A column increases the alcohol content of the finished product be removing most of the water from the still vapors. It does this by condensing the water (Fig. 6-3).

The column typically consists of a long vertical pipe filled with a series of baffles or filled with a loose packing material such as marbles, broken glass or metal shavings. The key to making the column work is to maintain the temperature within the column at a point which will keep the alcohol vaporized but not the water.

As the vapors from the still enter the column, they come in contact with the large surface area provided by the baffles or packing material. Since these surfaces are somewhat cooler than the surrounding air, some of the alcohol and water will condense on the surface as tiny droplets. If the column has been properly constructed and insulated so that the inside temperature stays above the boiling point of alcohol but not

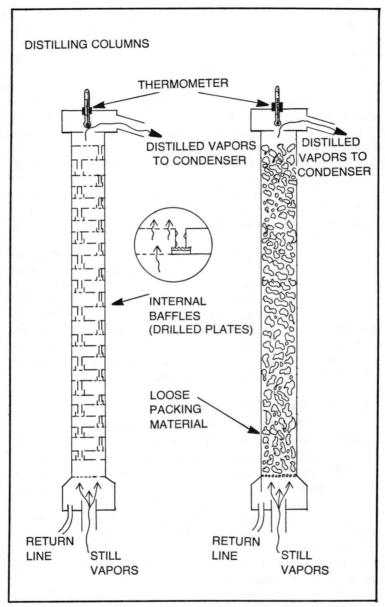

DISTILLING COLUMNS

THERMOMETER

DISTILLED VAPORS TO CONDENSER

DISTILLED VAPORS TO CONDENSER

INTERNAL BAFFLES (DRILLED PLATES)

LOOSE PACKING MATERIAL

RETURN LINE   STILL VAPORS

RETURN LINE   STILL VAPORS

Fig. 6-3. The column of the left is sometimes called a "rectifying column" and might contain 30 or more compartments. Height can range from 25 feet to several stories depending on the diameter. The column on the right is a "reflux" and can be 5 to 10 feet high.

water, the alcohol will re-evaporate and continue upward. The water remains behind and trickles to the bottom. This process happens over and over as the vapors rise through the column until most of the water has been removed and 170 to 190 proof alcohol vapor emerges from the top.

The most important factors in column operation are having a large surface area for these condensation/evaporation reactions to take place and maintaining proper temperature control.

A temperature gradient will exist within the column. It will be hottest where the vapors enter the column and coolest toward the top. The idea is to maintain a temperature of 173 degrees to 175 degrees F at the top so only relatively pure alcohol vapor will survive the journey. A cooler temperature than this will cause the alcohol vapors to condense within the column and remix with the water. Hotter than this will allow excessive amounts of water vapor to pass through.

Obviously, what comes out the top of the column depends on what goes in the bottom. Towards the beginning of the run, the vapors entering the column will contain a high proportion of alcohol. This means the column will do a relatively efficient job of removing the water to produce a high grade fuel. But toward the end of the run, the still vapors will be mostly water. Even though the column will remove much of the excess, the small amount of alcohol compared to the large volume of water will result in a lower proof product. In other words, the proof of the fuel coming out of the column will start out fairly high and taper off toward the end of the run. Because of this, the still should be shut down as soon as most of the alcohol has been recovered—just as with simple distillation.

To improve the overall efficiency of distillation or to produce a higher proof fuel, two and sometimes three columns are used in series. In such multiple column arrangements, the first column is operated at a higher than normal temperature (190 degrees F to 195 degrees F) to remove the bulk of the water. The second and third columns do the final job of purification. The finished product will usually be the highest possible with distillation, and that is 95 percent alcohol or 190 proof. To remove the remaining 5 percent water requires chemical drying.

The first still in such a multiple still setup can be referred to as the beer column or the stripping column. This first column could also employ a somewhat different approach for vaporizing the beer. A steam boiler can be used to feed hot steam into the bottom of the boiler at temperatures as high as 220 degrees F. The beer is pre-heated (but not boiling) and pumped into the column about halfway up. The liquid beer meets the hot steam, vaporizes and continues upward. This speeds the distillation process but also requires the building of a steam boiler instead of a still. Although this technique could be used in a small-scale alcohol plant quite effectively, the average do-it-yourselfer would probably be better off starting out with a simple still and single column arrangement.

Condenser

As the hot alcohol vapors emerge from the top of the column, some means of cooling is needed to condense them into liquid. If the vapors are routed through a long length of curved pipe, enough heat will be lost to convert most of the vapor into liquid. If that same length of coiled tubing is submerged in cold water, the vapors will condense much more quickly. Some means of rapid cooling is a must wherever large quantities of alcohol vapor are being generated (Fig. 6-4).

A simple condenser can be made by coiling a length of copper tubing. The reason for winding the tubing is to keep the vapors bouncing off the inside surface as they wind their way through the pipe. If it were straight, some of the vapor would probably flow on through without cooling sufficiently to condense.

As the vapors are cooled within the condenser, the surrounding air or water absorbs the heat. If the condenser is surrounded by a water jacket or submerged in a water tank, the water will begin to retain heat. This means a steady flow of water through the tank must be provided to carry away the excess heat and keep the condenser working at peak efficiency.

FUEL DRYING

The highest grade fuel you can achieve through distillation is about 95 percent (or about 190 proof). No matter how many columns you link in series, you can't remove that last bit of water. Once the concentration of alcohol reaches 95.6 percent, it is physically impossible to remove the remaining 4.4 percent water by boiling. The water and alcohol form a mixture known as an *azetrope*. The boiling point of the mixture is nearly identical to that of pure ethanol.

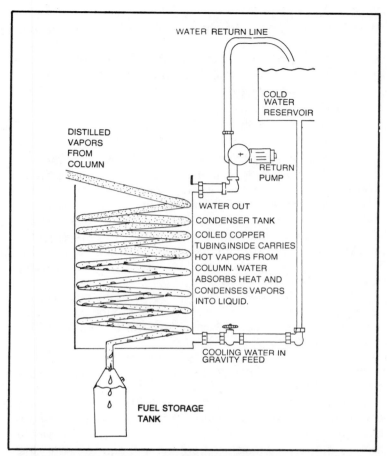

Fig. 6-4. A typical setup for a distillation condenser. Cooling water is recycled from the condenser to the reservoir.

So what do you do? Nothing, if you're burning straight alcohol fuel in your vehicle or tractor. Up to 20 percent water in the fuel won't make any difference other than to reduce the energy content per gallon. In fact, most experts agree that about 5 percent to 10 percent water in the alcohol fuel actually improves performance over pure alcohol.

The only time you need pure alcohol is if you're planning to mix it with gasoline to make gasohol. If more than about 1 percent or 2 percent water is present in an alcohol/gasoline blend, the two fuels will separate into distinct layers. This can cause all kinds of problems especially in cold weather. However, if you're burning straight alcohol fuel, you won't have to worry.

For those who do want to "dehydrate" their fuel to remove excess water or increase the purity, there are several techniques to choose from:

—Absorption.
—Dehydration with lime.
—Solvent extraction.

Absorption dehydration is relatively new and involves using some kind of material that absorbs water and not ethanol. The alcohol vapors are run through a column loosely packed with an absorptive material. This can be a second column in series with the first, a smaller column or section atop the main distilling column, or a separate column entirely. This type of column requires a higher operating temperature—around 180 degrees F to 200 degrees F. After the vapors pass through, they are cooled and condensed as usual.

Researchers have found a number of substances that will work in this respect, including cornstarch, ground corn kernels and shredded cellulose. Table 6-1 shows how some of the materials compare.

After each run, the material must be heated to dry or allowed to sit until the moisture evaporates. A possibility in the case of cracked corn kernels would be to dump the kernels and use them in the mash. An ingenious do-it-yourselfer might even come up with a way to circulate the distilled alcohol vapors through the ground corn before it goes in the mash. This would remove the excess moisture from the fuel and "pre-wet" the corn. A little extra moisture won't make a difference in the mash.

Table 6-1. Experimental Results of Alcohol Dehydration.

material	ethanol (%)	
	before	after
Cornstarch	73.7	99.0
Ground corn	77.0	97.7
Avicel (microcrystalline cellulose)	88.9	98.6
Corn residue	85.2	92.0
Sodium hydroxide	80.7	97.6
Buckeye CM cellulose	84.8	99.8
$CaSO_4$	90.1	98.0

The second technique for drying the fuel is dehydration with lime (calcium oxide or CaO). The liquid fuel is poured into a tank containing a small amount of lime. The lime reacts with any water present to form an insoluble compound known as calcium hydroxide. The unused lime and calcium hydroxide settle to the bottom and the purified alcohol fuel is drained off.

The trouble with this simple technique is that it is slow and cumbersome. For one thing, it takes about 35 pounds of lime for each gallon of water to be removed. For 50 gallons of 190 proof fuel, it would take 87.5 pounds of lime to remove the 2.5 gallons of water. The alcohol and lime must be stirred occasionally and even then it takes 12 to 24 hours for the reaction to be complete. The container must also be covered to prevent the alcohol from absorbing moisture from the air.

The calcium hydroxide that remains in the bottom of the tank can be disposed of, since lime is relatively cheap, or heated to 320 degrees F to 350 degrees F to regenerate it into lime. The high temperature causes the chemical reaction to reverse, changing the calcium hydroxide into calcium oxide. The water quickly evaporates.

The third technique for dehydrating fuel is solvent extraction. Commercial plants use various chemicals, such as benzene to remove water, but these chemicals are either too expensive, too toxic or too complicated to work with for the do-it-yourselfer. One water absorbing solvent that is practical for the do-it-yourselfer is regular or unleaded gasoline.

With this technique, you will be making gasohol by mixing gasoline and alcohol together. The normal proportion is 9 parts gasoline to 1 part alcohol. You can use as much alcohol in the blend as you prefer, but over 20 percent alcohol will require some carburetor modifications. You will achieve a "solvent extraction" effect of sorts because the gasoline will blend with the alcohol but very little water.

After the fuels are mixed and allowed to settle, the gasoline/alcohol and alcohol/water will begin to separate into distinct layers if there is more than 1 percent or 2 percent water present. The top layer will contain the gasoline/alcohol mix and the bottom layer alcohol and water. The gasohol can then be drained off and used as fuel. The remaining water/alcohol can be redistilled to remove the excess moisture.

FUEL STORAGE

As the alcohol is distilled and condensed, it should be routed into a covered storage tank. The tank *must* be vented to allow pressure within the system to escape. Otherwise, internal pressure can build to explosive proportions. Having no place to go, it will eventually make a vent of its own—rupturing pipes, fittings, or worst of all, the still itself. Remember, the still is like a giant tea kettle. All that steam and vapor has to go somewhere. Condensing reduces much of the pressure, but enough remains to do damage if it isn't treated with respect.

Another reason for covering the fuel tank (except to allow for adequate venting) is to minimize fuel loss through evaporation. Alcohol left exposed to the open air quickly disappears, just like gasoline. It also absorbs airborne moisture. On a particularly humid day, the alcohol might absorb enough moisture to actually lower the proof a noticeable amount.

For long term storage, fuel tanks must be sealed tight. Treat alcohol fuel as you would gasoline or any other flammable liquid. Even though alcohol is not as explosive as gasoline, it's still included in the same class of hazardous liquids as gasoline by most state and local fire regulations. This means that local ordinances might have a say as to where and how you store the fuel. Most laws require that fuel storage tanks be out of doors or inside fireproof buildings. Outside tanks usually must be located a minimum distance from buildings.

DENATURIZING FUEL

By law, alcohol fuel must be made unfit for human consumption by adding a small amount of gasoline, kerosene or ketone in accordance with U.S. Bureau of Alcohol, Tobacco & Firearms regulations. Be sure you clearly mark the alcohol fuel tanks with warnings that the fuel is unfit for drinking. It's also wise to keep your fuel storage tanks under lock and key to prevent accidental usage or pilferage.

TESTING PROOF

You should have some means of testing the proof of your fuel both during and after distillation. One way is to see whether or not a few drops of the stuff will burn. If it does, it's at least 75 proof to 80 proof (35 percent to 40 percent alcohol).

To accurately determine the worth of the fuel, use a good proof hydrometer (Fig. 6-5). A proof hydrometer works just like a sugar hydrometer except that it is calibrated to read liquids that are lighter than water.

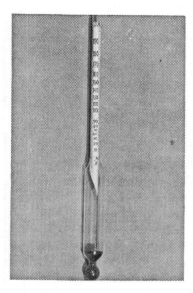

Fig. 6-5. A proof hydrometer. The instrument is placed in a sample of alcohol and read where it floats. This will show the strength or percentage of alcohol in the fuel. This particular hydrometer is available from the Mother Earth News.

To test the proof, place the hydrometer in a small sample and read the mark where it floats. If it's 160 or higher, you have fuel grade alcohol (Table 6-2). If not, redistill it to remove some of the excess water.

By periodically checking small samples of fuel while the still is running, you can keep an eye on the efficiency of the distilling operation. You will see a gradual decrease in the proof during a run as the time passes. This is because most of the alcohol boils off first. Once the proof begins to drop much below 140, you're starting to get into the tails. This alcohol should be collected in a separate tank for redistillation to remove excess water. If you run it in with the rest of the fuel, you will dilute the entire tankfull with excess water.

Low proof fuel coming out of the still during the first part of the run is also an indication of trouble. If the temperature at the top of the column is too high (much over 173 degrees F to 175 degrees F), an excessive amount of water will pass through. This means you either need more fire under the beer boiler or more insulation around the column.

Low proof might also indicate a low alcohol content in the beer. You can use a proof hydrometer to check the beer after fermentation to see if it's between 8 percent and 12 percent. If it's much less than 5 percent, you've got fermentation problems. Check the mash recipe to see if you did everything correctly.

Table 6-2. Percentage of Alcohol and Proof.

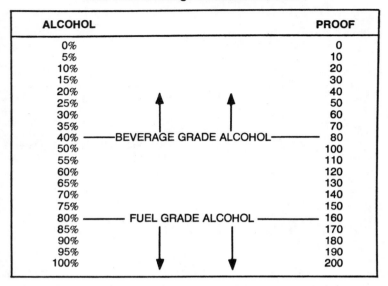

ALCOHOL	PROOF
0%	0
5%	10
10%	20
15%	30
20%	40
25%	50
30%	60
35%	70
40% ———BEVERAGE GRADE ALCOHOL———	80
50%	100
55%	110
60%	120
65%	130
70%	140
75%	150
80% ——— FUEL GRADE ALCOHOL ———	160
85%	170
90%	180
95%	190
100%	200

Low proof fuel or no proof fuel means you have boiled all the alcohol out of the beer in your still. Once the alcohol is gone, you are distilling water.

STILL OPERATION

The still should be filled with beer to two-thirds capacity. This leaves some room at the top for the vapors to expand and flow to the column. As soon as the beer is brought to a boil, the vapors will begin to circulate through the column. However, it takes some time to get things warmed up to their proper operating temperature. The time will vary from one still design to the next, but generally figure on 15 minutes to 30 minutes.

After the temperatures within the column begin to stabilize, the temperature at the top should be between 173 degrees F and 175 degrees F. If it isn't, then consult the troubleshooting guide that follows.

Temperatures Below 173 Degrees F at the Top of the Column

The problem is that not enough hot vapor is reaching the top of the column to maintain the desired temperature. Possible causes and cures include:

CAUSES	CURES
Insufficient vapor being generated inside the still.	Apply more heat to the still.
Excessive heat loss from the column to the surrounding air.	Add external insulation to the column and/or still. Wrap the lower third or half of the still with fiberglass batting to reduce heat loss.
Excessive heat loss inside the column.	The column is too tall. Shorten the height to reduce heat loss.
Excessive restriction inside the column.	Pack the column with coarser material or more loosely, or open up the baffles somewhat.

Temperatures above 175 Degrees F at the Top of the Column

The problem is that too much hot vapor is reaching the top of the column. Possible causes and cures include:

CAUSES	CURES
Too much vapor being generated inside still.	Reduce the heat under the still.
Insufficient heat loss from the column to the surrounding air.	Remove external column insulation, lengthen the column or increase restriction by using closer packing, coarser packing or increased baffling.
The column is too small for the still.	Install a larger column capable of handling the volume of vapor the still generates.

The idea is to achieve the right balance of vapor flow and heat loss. If the column is properly sized to the still, it only takes a little fine tuning to achieve this balance. After that, the operation will be relatively simple.

By watching the temperature at the top of the distilling column, you will notice a drop when all the alcohol has been boiled out of the beer. At the same time, the amount of fuel dripping out of the condenser will slow to a halt. Now is the time to shut things down. If the still is allowed to run, the temperature will drop for a short time and then rise to a point which will permit water to pass through. Distilled water is something you don't want. Shut the still down when you see the temperature drop.

You might encounter a false temperature drop. This happens when insufficient hot vapor reaches the top of the column—usually due to insufficient heat under the still. Whatever the cause, don't be fooled into thinking you're done. One way to tell a false drop from the real thing is through experience. After you've made a few runs, you'll have a rough idea about how long it takes to complete a run. If the still has been running only a fraction of its normal time, it's a sure signal there's more to come. Stoke up the fire and keep it going. You should also have a rough idea of how much alcohol you've recovered. If your still holds 50 gallons and you're expecting 4 to 5 gallons of fuel and you only have 1 or 2 gallons—keep going. Another check is to test the proof of the last fuel that came through. If it's still fairly high, then you're not into the end of the tails yet.

When you are through, douse the fire under the still. As soon as the stillage stops boiling, it can be drained out. If you're going to run another batch of beer through the still, just reduce the heat enough to stop the boiling, drain the still and refill with fresh beer.

STILL HEAT

To hold your production costs to a bare minimum, you should use the most inexpensive source of heat available. Many do-it-yourselfers can use wood, trash or crop residues such as stalks or cobs. Another source of inexpensive fuel is waste oil from the crankcase or automatic transmission of a motor vehicle. You could use fuel oil, propane, natural gas or even electricity if you prefer, but the costs of these fuels would make do-it-yourself fuel production an expensive hobby.

For wood, coal, charcoal or other solid fuels, use a draft control and door on the firebox. This will maximize the amount of heat generated by the fire. There are many wood burning stoves currently on the market that offer slow burning, efficient combustion. You might copy one of these designs or simply build your still over or around one of these wood burners. Whatever your choice, make sure the firebox is a safe distance from the fuel storage tanks or any plumbing that might spill alcohol near the fire.

VACUUM DISTILLATION

A special technique that can significantly reduce the amount of heat required to "fire" the still is a process known as *vacuum distillation*. To understand how vacuum distillation works, you must first understand that temperature and pressure are directly related (Fig. 6-6, 6-7 and 6-8). Change one and you affect the other. For example, an increase in pressure increases the temperature at which a liquid boils. This is what a pressure cooker does. The pressure squeezes the liquids so that the bubbles can't form as quickly.

By the same line of reasoning, reducing the pressure lowers the temperature at which the liquid will boil. The pressure is taken off the liquid so that the bubbles can form more quickly. This is what happens at a high altitude when water is boiled. Because the air is thinner at the higher altitude, the pressure on the water is less than the normal 14.7 pounds per square inch at sea level. The water will boil at a lower temperature, perhaps 195 degrees F instead of 212 degrees F.

With vacuum distillation, a vacuum pump is used to reduce pressure within the system so that the beer in the still is easier to boil. If the pressure is reduced enough, the beer might begin to boil if the outside temperature is 70 degrees F or higher. This eliminates the need for a fire altogether.

To make vacuum distillation work, the still, column, condenser and fuel receiving tank must be air tight. A vacuum pump is connected to the system to pump out air and reduce the internal pressure. The equipment must be constructed of a heavy enough material to withstand the partial vacuum without collapsing. After the desired degree of vacuum is created within the system, the pump is switched off. A small amount of heat is applied to the still to get things started. It will then run as usual. The only difference will be that the temperature at the top of the column will be reduced by an amount proportional to the amount of vacuum within the system.

After distillation is complete, the system is vented and allowed to return to normal pressure. This also stops the beer from boiling just as dousing the fire does for normal distillation.

The thing to remember about vacuum distillation is that the greater the vacuum created within the system, the lower will be the boiling point of the liquid in the still. The rate of boiling can be controlled by varying the vacuum—just as varying the amount of heat will cause the beer to boil faster or slower.

On a warm day, solar energy would be sufficient to power the still, assuming a low enough vacuum existed within the system. Painting the outer surface of the still black, encasing it in a transparent outer covering or using reflectors to direct more light at the still would help enhance the absorption of solar radiation.

SOLAR DISTILLATION

Unless combined with vacuum distillation, solar distillation

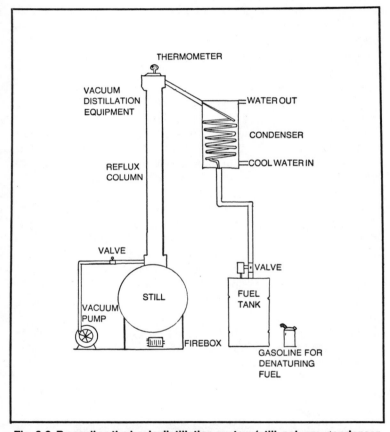

Fig. 6-6. By sealing the basic distillation system (still, column, condenser and fuel tank) to the outside atmosphere and connecting a vacuum pump, the internal pressure can be pumped down to significantly lower the boiling point of the beer in the still. This reduces the energy needed to "fire" the still. If a high enough vacuum is created within the system, the beer will begin to boil if the outside temperature is 70 degrees F or higher. In operation, the vacuum pump is allowed to run until the desired reduction in pressure is reached. Then it is shut off and the valve linking it to the system closed. After distillation is complete, the valve is again opened and the system allowed to return to normal pressure.

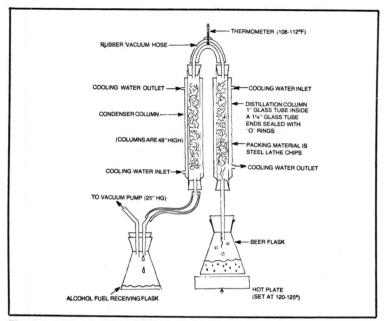

Fig. 6-7. A simple laboratory vacuum still.

is slow and inefficient. Although a simple solar collector can be built and filled with beer, controlling the temperature within the collector can be troublesome. If it gets too hot, you will boil off a lot of steam with the alcohol. If it doesn't get warm enough, it'll take a great deal of time to evaporate all the alcohol.

Solar stills do work and they can be used to make alcohol fuel. But unless they are combined with a traditional distilling column, the proof of the finished product will be quite low. In addition, unless they are combined with some type of vacuum system, they are apt to be painfully slow. A typical output might be a gallon of alcohol fuel per day for a simple solar still.

The biggest advantage of the solar still, however, is the energy source. For practical purposes, the energy is free. Even on a cloudy day, a properly built solar collector will obtain enough heat gain to distill some alcohol.

One application for solar distillation that seems quite promising is that of "pre-distillation". The beer from the fermenter is "concentrated" by running it through a simple solar still to remove a portion of the excess water. The beer evaporates and condenses, leaving some of the water behind. If this concentrated beer is then run through a regular still, the amount of

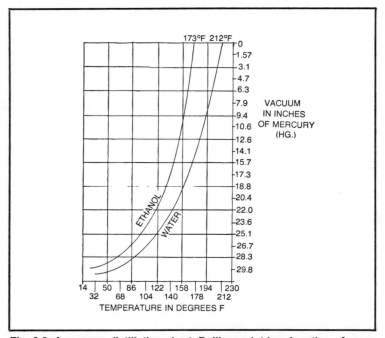

Fig. 6-8. A vacuum distillation chart. Boiling point is a function of pressure. Reducing the pressure within the still with a vacuum pump lowers the boiling point of the beer. This reduces the amount of heat energy needed to power the still, which in turn reduces operating and production costs. The actual boiling temperature of the beer will fall somewhere between the two lines at a given pressure depending on the relative proportions of alcohol and water. To gain a significant reduction in the boiling temperature means pumping the system down to at least 22 inches of mercury. This also means the distilling equipment must be of heavy enough construction so it won't collapse under outside atmospheric pressure.

energy needed to make the final distillation will be reduced by one-third to one-half.

If the beer is 90 percent water and 10 percent alcohol before it enters the solar still, it can be from 25 percent to 50 percent alcohol when it emerges. As you can see, the energy needed to fire the regular still will be greatly reduced.

Such a system might be set up so that beer is fed directly from the fermenter to the solar still on a continuous basis. Once enough concentrated beer is collected for final distillation, the regular still can be fired up and put into use.

Chapter 7
Basic Stills

Regardless of the distilling technique you use—simple, reflux, vacuum, solar or solar vacuum—workmanship and safety are very important considerations that should be foremost in your mind as you design and build a still.

If you're serious about fuel production and want to build a still, then build it right and build it to last. Using common sense during the construction of the still will result in a unit that is both safe and reliable to operate. Nothing is more irritating than to have an equipment failure in the middle of a run. Sure, every new system will have a few bugs, but many problems are preventable. A pipe joint hastily assembled will undoubtedly develop a leak. A poorly sealed seam might suddenly burst. A column or chimney that is inadequately braced might topple over. Such mishaps are not frustrating but they are also potentially dangerous. Alcohol is a flammable liquid if it happens to splash against a hot surface it could easy ignite.

Shoddy workmanship, shortcuts, or carelessness in construction or operation of the still can transform an innocent looking collection of tanks and tubes into a lethal bomb. The purpose of this book is to help you make your own fuel—not your funeral—so please, use your head and heed the precautions described in this and other chapters.

PLUMBING DETAILS

Columns, chimneys and any length of pipe much over 4 or 5 feet in length should be braced for support. The heat and vibration that is produced during operation will take its toll on your equipment if pipes are not adequately supported. The braces must be rigid enough to hold things in place yet be flexible enough to allow for heat expansion.

All threaded joints must be sealed with pipe compound to prevent leaks. All joints between pipe flanges and tanks must be sealed with a cork or rubber gasket, or with a gasket material such as silicone or RTV sealant. If you're in doubt as to how to fit pipes or make leak proof connections, refer to TAB book No. 914, *The Home Plumber's Bible* or No. 1214, *Plumbing With Plastic*. A still is like a giant tea kettle in many respects except that the pressures generated are much greater. Normally the pressure doesn't have a chance to build up because the vapors are vented through the column. If something should happen to restrict the flow of vapors (the column becomes clogged) or if the column can't handle the volume of vapors being produced within the still (the still is too hot or the column too small) you and your still could be in for trouble.

To avoid mishaps, all stills should incorporate some type of safety pressure release (Fig. 7-1) on the beer heating tank. If a problem develops and internal pressure exceeds about 10 psi, the safety valve pops open and vents the pressure before it can do any damage. Such safety devices can be purchased from a variety of sources or you can make your own.

A simple pressure release can be made as part of the filler cap. A rubber or cork plug, or an oversized rubber ball, can be used as a stopper to plug the filler hole. In the event of a pressure back up, the plug will blow out of its hole allowing the pressure to vent harmlessly into the atmosphere. If this happens, your first reaction should be to extinguish the fire and get the system under control.

A cap that is screwed on a tight fitting filler cap should never be used on a still unless other means of venting excess pressure are provided. Naturally, the still must be sealed to prevent the loss of alcohol vapors during normal operation, but it must also have a safety valve in case something goes wrong.

An auxiliary safety vent can be made by welding or brazing a short length of 2-inch or 2½-inch pipe to the top of the still. The pipe extends through the top of the beer tank like a small chimney. The top of this pipe is then plugged with a rubber or cork stopper, or a rubber ball slightly larger in diameter than the inside diameter of the pipe. If pressure builds up, it will blow out the plug.

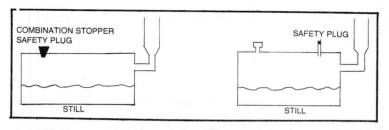

Fig. 7-1. Safety valves should be installed atop the still as shown. The valve on the left consists of a rubber, cork or wood plug that fits snuggly into a hole cut in the top of the still. If pressure within the still exceeds about 10 psi, it will blow out the plug and vent harmlessly into the atmosphere. Because of this, the plug should not fit too tightly. The safety valve at right works on the same principle but consists of a length of pipe plugged with a rubber ball. The ball should be about one-quarter inch larger in diameter than the inside diameter of the pipe.

FIRE CONTROL

As I have warned repeatedly, a sealed tank partially filled with liquid and with a fire underneath is a potential bomb. Many a moonshiner has bit the dust because somebody failed to keep an eye on the fire under the still. Crude stills typically have no provisions for emergency pressure release—except to explode if the brew overheats.

You should have close at hand some means of rapidly extinguishing the fire. A bucket of water could do the trick or you could build a water line and sprinkler head into the top of the firebox for rapid shutdown. A twist of a valve would flood the fire with water. An air or damper control will also work, but the results are much slower. In an emergency, the quenching might come too late. For gas or oil fired stills, a simple fuel shutoff valve should be located nearby. The same applies to an electrical heating element.

The still should be far enough away from fuel pipes or tanks to minimize the risk of fire if alcohol is accidentally spilled. The firebox or burners should be several inches off the floor or built on a raised platform so that any fuel spilled on the ground won't have a direct access to flame.

VENTILATION

If the still or fermentation tanks are located inside a building, garage or shed—assuming local zoning laws and fire ordinances permit it—be sure to provide adequate ventilation for the firebox and for CO_2 escaping from the fermenter. Fuel tanks must be covered since alcohol evaporates rather quickly.

SAFETY

Buildings must have all electrical wiring routed through airtight conduit and junction boxes and all light bulbs enclosed in airtight glass housings. This prevents a spark from igniting any alcohol vapors that might have accumulated because of leaks, evaporation or carelessness. Another suggestion to help make fuel production safe is to paint the plumbing with a color code according to what the pipes carry. For example, you might paint water lines blue, beer lines yellow and alcohol lines red. Labeling pipes, valves and other components helps to prevent mixups. To remind you of what is going on inside all that plumbing, you might paint arrows on the pipes showing the direction of flow.

CONSTRUCTION MATERIALS

The materials needed to build a basic still are detailed below.

Beer Still. It should be a metal tank of rust and corrosion resistant material. Stainless steel is one of the best materials, but unfortunately it is quite expensive. The same applies to copper. Ordinary steel or recycled steel drums, tanks or kegs are more affordable and will work fine—as long as the bare metal is painted or covered with a protective coating such as fiberglass, epoxy paint or hard enamel. Recycled aluminum beer kegs will also work and require no special preparation. However, plumbing fixtures might have to be bolted to the tank with flanges since aluminum requires special techniques to weld or solder. Vacuum stills must be strong enough to withstand outside atmospheric pressures that would collapse an ordinary still. This means using a heavy gauge steel or a recycled pressure tank such as an old propane or low pressure tank. Recycled beer kegs will also work, but steel drums will probably require additional internal bracing.

Plumbing. Copper tubing, steel pipe, cast iron or plastic pipe can be used. For plastic pipes that carry hot vapors or liquids, check the manufacturer's recommendations for temperature tolerance. Most plastic pipes can handle boiling water without softening or melting. All pipe joints, metal or plastic, must be sealed with some types of gasket cement or sealer.

Firebox. An open fire is not recommended. Use only an enclosed fire, electrical heating element, enclosed gas or oil burner, or solar energy to provide heat. For many do-it-yourselfers, a metal or brick enclosed firebox with air and damper controls offers the most economical and practical means of heat. This will enable you to burn wood, trash, cobs, leaves, stalks, coal, charcoal or even manure to fire the still. Another alternative fuel source is waste oil from vehicles or farm equipment. Burn it through a modified oil burner.

Methane gas from a compost heap can also be used through an ordinary gas burner. So-called "waste" heat can be another alternative. You might try recycling heat from a nearby furnace, chimney or exhaust heat to power your still.

Fuel Storage Tanks. All tanks must be metal, fiberglass or plastic, and clearly labeled "Alcohol Fuel" to avoid mixups. Glass or cermaic jugs should not be used because of the possibility of breakage. It is also illegtal in most areas to store or transport a flammable liquid in a breakable container.

Pumps. There are many types of pumps available for circulating liquids or moving liquids from one tank to another. A water cooled condenser will undoubtedly require some type of return pump on the water line. Pumps are expensive so plan your equipment layout to make the best use of gravity. Do-it-yourself pumps can be made by recycling an old washing machine or dishwasher pump.

Thermometers. Since alcohol production temperatures are relatively high, an ordinary house thermometer won't work unless you're using high vacuum distallation. You will need a thermometer that is designed to read at least as high as 200 degrees F and preferably 250 degrees F. Such instruments can be purchased from laboratory supply houses. Less expensive alternative include candy, meat or oven thermometers. However, these might not be sufficiently accurate. Thermocouples with remote meters are also available to take the effort out of temperature monitoring. Instead of climbing to the top of the column to check the temperature, you mount the meter with any other controls in a convenient spot.

BASIC DESIGNS

Choosing a design involves a number of considerations, including cost, size and output. To help you decide, the main

Table 7-1. Comparison of Basic Still Designs.

Type of Still	Equipment	Advantages	Disadvantages
SIMPLE STILL	large kettle, pot or tank for boiling beer heat source (firebox, burner, etc.) safety valve high temperature thermometer coil condenser (air or water cooled)	easy to build (no column used) inexpensive to construct simple to operate	low proof fuel requires several redistillations to raise the alcohol content to fuel grade level
REFLUX STILL	large tank to boil beer heat source (firebox, burner, etc.) safety valve reflux column high temperature thermometer at top of a column water cooled condenser fuel receiving tank	efficient operation: makes fuel grade alcohol first run through the still	requires the addition of a column more difficult to build than a simple still more difficult to operate than a simple still
VACUUM REFLUX STILL	large tank to boil the beer (must be strong) heat source—requires much lower heat than an ordinary still or no heat at all if the vacuum is high enough thermometer—lower temperatures mean than an ordinary household thermometer might work reflux comumn (might require water cooling) water cooled condenser fuel receiving tank (airtight) vacuum pump vacuum gauge	requires less heat input to distill fuel makes fuel grade alcohol first time through efficient operation	requires heavier construction to withstand external pressures more difficult to build than a simple still more difficult to operate than a simple still
SIMPLE SOLAR STILL	solar collector to heat beer beer feed line to collector hi-temperature thermometer fuel receiving tank	low cost to operate (solar energy) simple to build simple to operate (in most cases)	typically slow output low proof fuel as with simple distillation requires redistillation to raise proof

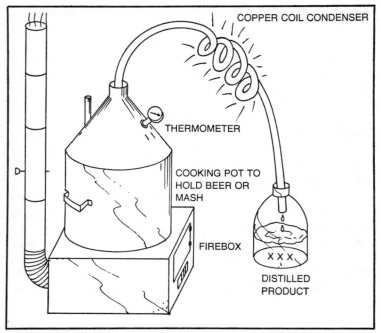

Fig. 7-2. After the mash has been fermented, the alcohol-containing beer is strained out and poured into the pot atop the firebox. The liquid is brought to a slow boil. The alcohol vapors boil off and flow through the copper coil where they radiate off heat and condense into liquid again. The distilled product is then run through the still again to increase the proof and remove harmful fusel oils.

components, advantages and disadvantages are listed in Table 7-1.

DISTILLATION REVIEW

The least sophisticated method of distillation is to boil the beer in a tank, collect the vapors and then recondense them into liquid. This one-step distillation process, unfortunately, doesn't yield fuel grade alcohol. At best, the condensed vapors will be 40 percent to 45 percent alcohol (80 to 90 proof). The rest will be water. For fuel, you need at least 80 percent alcohol (160 proof).

The first vapors to boil off, called the heads, contain the highest portion of alcohol while those toward the end, called the tails, will be mostly water. To increase the concentration of alcohol and remove the excess water, the fuel must be redistilled repeatedly. If you are using simple distillation or simple solar distillation, you are wasting time and energy if your heat source is something other than the sun. Because of this, simple distillation is not recommended for the serious fuel producer. Instead, you should think along the lines of a reflux still, a vacuum still or a solar powered vacuum reflux still.

Figure 7-2 shows a simple "moonshine" still that could be used for fuel production on a limited basis (though not recommended). The large kettle is filled to about the three-quarter level with either beer or mash. The kettle is then heated until the contents begin to boil. The vapors rise and pass through the *worm*, a long length of coiled copper tubing that serves as the condenser. The vapors lose heat, condense and drip out the end into the receiving flask. A non-breakable metal or plastic container should be used to collect the liquid.

Using such a still to make moonshine instead of fuel could lead to real problems. Besides breaking the law, the moonshiner (and his victims) runs the risk of lead poisoning. Even where expensive copper tubing is used, soldered pipe joints and fittings can react with the alcohol and water vapors to produce lead salts. It is impossible to see or taste lead salts and only a very small amount is needed to blind, paralyze or kill a human being. Lead accumulates in the drinker's body like arsenic. It continues to build up until serious stomach disturbances result, and ultimately brain damage and death. That is part of the reason the government is so hard on moonshiners. Another reason is the tax revenues of $10.50 per 100 proof gallon for drinking alcohol.

THE REFLUX STILL

Essentially the same thing as a simple still with a column attached to it, the reflux still (Fig. 7-3) is the fuel producer's choice. In effect, the column makes many redistillations during a single run which results in a much higher concentration of alcohol in the finished product. This saves time and reduces production costs.

The column is designed to expose a large surface area to the rising vapors within so that the vapors will condense and re-evaporate as they travel toward the top. By controlling the temperature within the column, most of the water will condense and relatively pure alcohol vapors will rise to the top. The vapors are then cooled, condensed and collected in a tank. The height of the column is determined by the diameter of the column and the amount of restriction within the column.

COLUMN SIZING

A small commercial column might measure 24 feet high, 12 inches in diameter (Fig. 7-4) and contain 40 or more reflux plates (baffles). These plates can be stacked 6 to 7 inches apart inside the column. The plates are drilled with three-sixteenth inch to three-eighths inch holes so that about 10 percent to 15 percent of the surface area will pass vapors. Such a column might produce 25 gallons of fuel per hour. Large commercial

Fig. 7-3. A reflux still built by the Mother Earth News.

Fig. 7-4. The column and condenser on a reflux still owned by the Zeith-amer family in Alexander, Minnesota. The column uses reflux plates rather than packing material.

columns might be several feet in diameter and be 5 or 6 stories tall. Fuel yields can be as much as 100 gallons per hour.

A practical column size for the beginning fuel producer might be a column with an inside diameter of 2 inches and a height of 4 feet. The column could be packed with marbles, broken glass or metal chips and shavings. A small column like this would produce about one-half to 1 gallon per hour.

A slightly larger column, 3½ inches in diameter and 10 feet tall, could yield up to 7 gallons of fuel per hour. Increasing the diameter to 4 inches would enable the 10-foot column to produce as much as 10 gallons per hour.

The actual length of the column is best determined experimentally after the still is constructed. Use the above figures as a reference point or refer to stills described elsewhere in this book. By watching the outlet temperature at the top of the still, you will be able to judge its performance. If little or no vapor is coming out the top or if the vapors are too cool, the column might be too tall, too large in diameter, too loosely packed or too big for the still. If, after some careful consideration, you believe that the column should be about the right size for the still, try insulating it to reduce the heat loss. Pack more material into the column or shortening the length of the column itself. If the column is constructed of bolted sections, fine tuning for maximum efficiency is much easier. Just add or subtract a section to modify the height.

The average do-it-yourselfer will probably find building a column with packing much easier than one with baffle plates. The restriction within a packed column depends on how coarse the packing material is—big marbles versus small marbles, large chips versus small chips,—and how density it is packed. With reflux baffles, the size and number of holes determines the amount of restriction. More holes or larger holes allow vapors to rise more quickly. Fewer holes or smaller holes slow vapors down. One technique to determine the right size and number of holes is to start by drilling enough three-sixteenth inch holes to open up about 8 percent of the surface area of the plates. The plates are then stacked about 6 inches apart inside the column. A trial run is made and the temperature noted. If there is too much restriction, either more holes should be drilled or the existing holes drilled to a larger size. Another alternative is to remove some of the plates. If there isn't enough restriction, then more plates need to be added and the height of the column extended.

FITTING THE COLUMN TO THE STILL

It's very important that the column capacity and still capacity be matched. Although there is a fair amount of leeway, a small still that holds 15 to 20 gallons won't work well with a column that is 4 inches in diameter and 10 feet tall. There's just too much column and not enough vapor. The reverse is also true. A 500 gallon still would quickly overload a 2-inch column that was perhaps only 4 or 5 feet tall. The vapor would flood the column and allow excessive amounts of water to spill over into the condenser.

Table 7-2 will serve as a rough guide to matching still size and column capacity.

Table 7-2. Column Size and Still Capacity.

still capacity	approximate column size	production rate
up to 25 gallons	2″ × 48″	½-gallon/hr.
25 to 100 gallons	3½″ × 10′	2 - 7 gallons/hr.
100 to 500 gallons	4″ × 10′	10 gallons/hr.

THE VACUUM STILL

The idea behind vacuum distillation is to lower the boiling point of the beer by reducing the internal pressure within the still. A vacuum pump is used to pump out most of the air. This will lower the boiling point of the beer from around 200 degrees F down to 120 degrees F or lower, depending on the degree of vacuum. The greater the vacuum, the lower the boiling point. If enough vacuum is put on the system, the beer will boil spontaneously at room temperature—eliminating the need for a heat source entirely. The big benefit here is reducing the cost of producing the fuel. By lowering the boiling point, solar energy becomes practical for reflux distillation. The same applies to waste heat or other low temperature heat sources.

The vacuum concept can be applied to a simple still or a reflux still, although the reflux still is preferred because of its higher proof fuel output.

The biggest problem with building a vacuum still is making it strong enough to resist the outside atmospheric pressure. It must also be airtight to hold the vacuum. Any leaks will quickly allow air to enter and destroy the vacuum.

The Vacuum Column

Because vacuum distillation involves much lower temperatures, the distilling column might have to be water cooled to achieve the desired condensation/evaporation cycle necessary to separate the excess water. This is because the temperature differential between the vapors inside the pipe and the surrounding air is not very great. Remember, the vapors entering the vacuum column might be only 125 degrees F or so compared to 70 degrees F to 85 degrees F for the outside air. Compare that to a normal atmospheric still that generates vapors at 200 degrees F. As you can see, there is a significant difference.

The Vacuum Condenser

The condenser on a vacuum still also requires some changes from a condenser on a normal reflux still. To get the vapors to recondense, more cooling effort must be put into the operation. This means a larger condenser with more surface area. One technique is to build a second column (lots of internal surface area) and use it as the condenser.

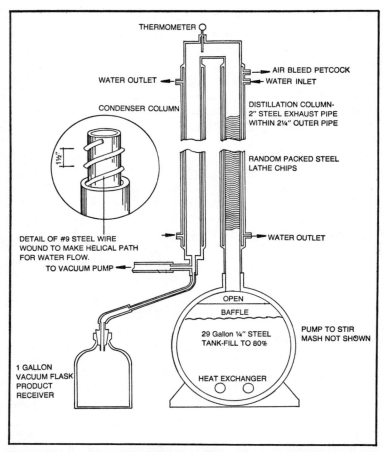

THERMOMETER

AIR BLEED PETCOCK
WATER INLET

WATER OUTLET

DISTILLATION COLUMN-
2" STEEL EXHAUST PIPE
WITHIN 2¼" OUTER PIPE

CONDENSER COLUMN

RANDOM PACKED STEEL
LATHE CHIPS

1½"

DETAIL OF #9 STEEL WIRE
WOUND TO MAKE HELICAL PATH
FOR WATER FLOW.

WATER OUTLET

TO VACUUM PUMP

OPEN

BAFFLE

29 Gallon ¼" STEEL
TANK-FILL TO 80%

PUMP TO STIR
MASH NOT SHOWN

1 GALLON
VACUUM FLASK
PRODUCT
RECEIVER

HEAT EXCHANGER

Fig. 7-5. The vacuum still must be constructed of heavy gauge steel (such as one-quarter inch) to withstand atmospheric pressure. A vacuum pump is attached to the system to produce a vacuum of around 25 inches of mercury. Because the system is air tight, the vacuum will hold once the pump is disconnected. The lowered pressure reduces the boiling point of the beer to around 120 degrees F to 125 degrees F, meaning an electrical heating element can be used to "fire" the still.

If the vacuum within the system is such that the vapors boil spontaneously at room temperature, then the condenser must cool the vapors *below* room temperature to recondense them. This means you will need a source of very cold water, a deep well, ice water or a refrigeration unit. Once the vapors are recondensed and collected in a storage tank, the tank must be kept below room temperature for the duration of the run to keep the fuel from re-evaporating. Of course, once the run is completed the system can be vented and allowed to return to normal pressure. This will keep the fuel you've made from boiling.

Figures 7-5, 7-6 and 7-7 show a solar powered vacuum still and two vacuum reflux stills. The operating conditions for all three stills is as follows:

—Vacuum: 25 inches of Mercury.
—Temperature of the beer: 122 degrees F.
—Cooling water inlet temperature: 60 degrees F.

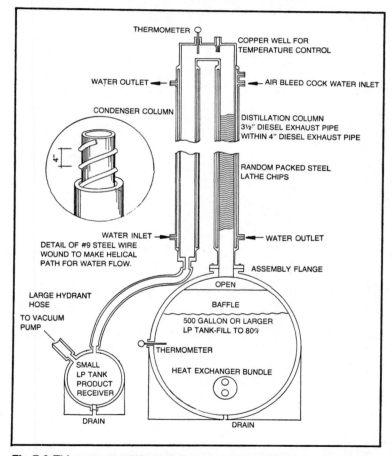

Fig. 7-6. This vacuum still is made from a recycled 500 gallon low pressure tank. An electrical heating element is used to heat the beer to a boil at approximately 120 degrees F to 125 degrees F at 25 inches of mercury. The capacity of this still is about 7 gallons of 190 proof fuel per hour.

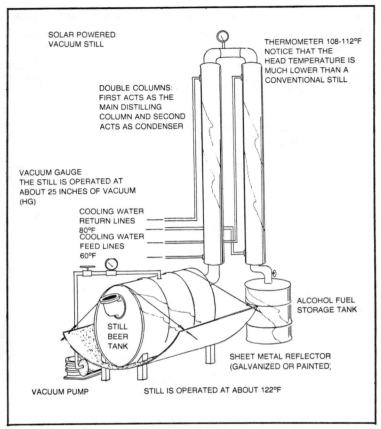

Fig. 7-7. With this system, a vacuum pump is used to create a partial vacuum. This lowers the boiling point of the beer so that solar heat can power the still. The still is painted black and sits atop a solar reflector.

—Vapor temperature at the head of the column: 108 degrees F to 112 degrees F.

—Btu input to distill alcohol: 10,000 Btu/gallon.

The still shown in Fig. 7-5 will produce about one-half gallon of 190 proof alcohol and the still shown in Fig. 7-6 will produce about 7 gallons of fuel per hour.

The larger still (Fig. 7-6) uses a recycled low pressure tank as the main beer tank and a recycled low pressure tank for receiving the fuel. These heavy tanks are needed to withstand outside atmospheric pressure that would probably collapse an ordinary steel drum.

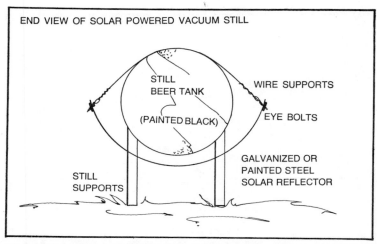

END VIEW OF SOLAR POWERED VACUUM STILL

STILL BEER TANK

(PAINTED BLACK)

WIRE SUPPORTS

EYE BOLTS

GALVANIZED OR PAINTED STEEL SOLAR REFLECTOR

STILL SUPPORTS

Fig. 7-8. To achieve maximum solar gain in heating the beer tank, the still tank is painted flat black and oriented on a north-south axis. The reflector can be galvanized steel or steel painted with white, aluminum or other reflective paint. Curvature of the reflector can be easily adjusted by lengthening or shortening the wire supports. Holes for the still supports must be cut in the reflector.

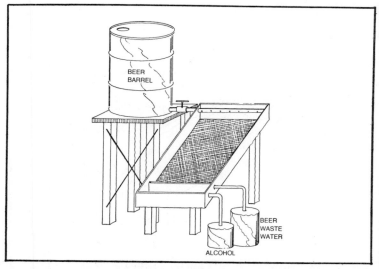

BEER BARREL

BEER WASTE WATER

ALCOHOL

Fig. 7-9. A typical setup for a solar still. Gravity feeds the beer into a sealed, glass-covered box. Beer dribbles out of feed pipe onto black burlap. Heat from the sun evaporates the alcohol. The vapor condenses on the glass and trickles down to the collection trough. The unevaporated water drains toward the bottom and is diverted out.

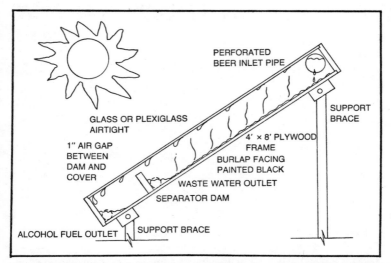

Fig. 7-10. Side view of a simple solar still. Beer is fed into an airtight collector box through a perforated pipe that runs the length of the top. The beer drips down onto the burlap covering the surface. Solar heats evaporates the alcohol which condenses on the glass and trickles to the bottom.

Fig. 7-11. The sun heats the beer in a tray inside the sealed box. Alcohol evaporates and condenses onto the glass. Drops trickle down into the collection trough and out. The remaining water is drained or dumped from the tray.

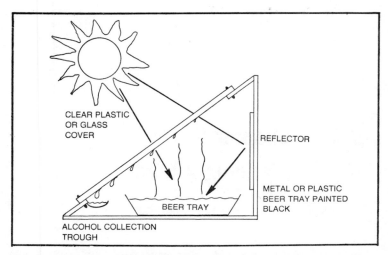

Fig. 7-12. Heat from the sun evaporates alcohol from the beer in the tray. Vapors rise and condense on the glass. For the process to work, the temperature within the solar box must be maintained between 175 degrees F and 180 degrees F.

SOLAR STILLS

The biggest advantage of the solar still (Fig. 7-7 through 7-12) is its heat source. The problem is that solar distillation is like simple distillation unless it is combined with a vacuum reflux column as described earlier. The resulting fuel contains considerable water. If you are in no hurry and don't mind making repeated distillations, solar heat will work on a limited basis.

The basic equipment consists of an airtight box of some sort with a large surface area to collect solar radiation. The beer is allowed to trickle into the collector panel and evaporate at a rate which will hopefully leave much of the water behind. By limiting the temperature, more alcohol than water will evaporate. The vapor then condenses on the relatively cool glass or plexiglass cover and trickles to the bottom. The drops collect and then run out and into a fuel receiving tank. The unevaporated water, meanwhile, continues its journey to a dam at the bottom of the collector and runs out an exit hole.

An open pan of beer might be considered a crude solar still. Over a period of time, most of the alcohol will evaporate along with some water. If the pan is inside some type of container or housing so that the vapors can be collected and condensed, simple distillation can be accomplished.

Chapter 8
Planning Ahead

Now that you've been introduced to the basics of fuel production, you're in a better position to determine whether or not alcohol fuel is practical for you. To make that decision, you need to consider the following:

—How much fuel per year do you need?
—How much effort are you willing to invest in making that fuel?
—What kind of equipment will be necessary?
—How much will it all cost?

Your annual fuel consumption depends on your particular set of circumstances. The fuel requirements for a farmer will depend on such things as the number of acres farmed, terrain and soil conditions, the type of crop planted and method of planting, the amount of tillage, fertilizer and cultivation required, weather conditions, the size and horsepower of the tractor and the type of planting, cultivating and harvesting equipment used. The variables are infinite, but on the average most farmers in the United States consume between 3 to 7 gallons of diesel fuel per acre per year. Fuel consumption for a city dweller depends on the miles driven per year, the type of driving (highway versus stop-and-go city driving) and the fuel efficiency of the vehicle.

FIGURING FUEL REQUIREMENTS

To estimate your annual fuel requirements, do one of the following:

- Total last year's fuel bills.
- Divide the total number of miles driven last year by your average miles per gallon.
- Multiply the total number of acres farmed by an estimated fuel consumption per acre. Figure 3 to 7 gallons of diesel fuel per acre per year, or 3.5 to 8.5 gallons for gasoline powered tractors.

To figure how many gallons of alcohol fuel per year you will need, multiply your estimated fuel requirement by a "fudge factor" of 1.20. A conversion to alcohol, might increase fuel consumption from zero percent to as much as 40 percent. Much depends on the type of conversion and modifications you make. This is because ethanol contains fewer Btu's per gallon than gasoline or diesel fuel. The 20 percent fudge factor is based on an average for most simple conversions such as drilling out the carburetor jets and advancing the ignition timing. More sophisticated conversions can actually produce an improvement in fuel consumption.

When you know how much alcohol you need to become energy self-sufficient in farming or transportation, you can begin to calculate the amount of raw material required, how big your still should be and how much time you'll need to make the fuel.

Example. On a farm with 600 acres of corn and a fuel consumption rate of 3600 gallons of diesel fuel per year, how much corn would it take to obtain energy self-sufficiency? To find the answer, first determine how many gallons of alcohol fuel are needed per year and then figure how much corn is necessary to make that amount of alcohol.

To calculate the amount of alcohol fuel needed per year, take 3600 gallons of diesel consumed per year times the 1.2 conversion factor. Answer: $3600 \times 1.2 = 4320$ gallons of alcohol fuel needed per year.

Divide the number of gallons of alcohol needed by the number of gallons of alcohol produced by a bushel of corn (refer to Table 3-2).

The yield from 1 bushel of corn is approximately 2.5 gallons of ethanol. Therefore, 4320 divided by 2.5 equals 1728 bushels of corn. To make enough alcohol for energy self-sufficiency, 1728 bushels of corn are needed.

If the average crop yield is 100 bushels of corn per acre, then approximately 17 or 18 acres of a 600 acre farm will be needed for fuel production. That figures to be about 3 percent of the potential crop—not bad for do-it-yourself fuel.

Most so-called alcohol fuel experts estimate that a typical farm can be energy self-sufficient by using about 5 percent of the grain or corn crop for fuel production. As you can see, the farmer in the example came in under that by a significant margin.

Example. For a compact car that averages approximately 22 miles per gallon and is driven 15,000 miles per year, how much alcohol fuel would it take to obtain energy self-sufficiency? To find the answer, first figure the total number of gallons of gas consumed in a year and then multiply the alcohol conversion factor of 1.2. Answer: 15,000 miles divided by 22 mpg equals 682 gallons of gas consumed per year. The conversion factor of 1.2 times 682 gallons of gas equals 818 gallons of alcohol fuel per year. To be energy self-sufficient, it takes 818 gallons of alcohol fuel per year.

FIGURING STILL SIZE

Based on the estimate for your yearly alcohol fuel requirement, you can calculate how large your still should be.

See Table 8-1 and Figure 8-1. Another consideration is how much time you will have available to spend making alcohol. If the farmer needs 4320 gallons of 180 proof alcohol fuel per year he has only enough spare time to make alcohol once a month, then his still capacity should be: 4320 gallons divided by 12 months equals 360 gallons.

Remember that the 360 gallons is 360 gallons of finished product—not 360 gallons of watered-down mash. It takes a lot of mash to make a little alcohol. With corn, for example, each bushel of ground corn is mixed with 32 gallons of water (4 parts water for each part solids). This means that 40 gallons of mash will yield about 2.5 gallons of alcohol (based on one bushel of corn equalling 2.5 gallons of ethanol). In other words, for every gallon of finished product, the fermenting and distilling equipment must handle 16 gallons of mash (based on corn). This number will vary with the type of crop being fermented.

The total capacity of the fermenting and distilling equipment based on one production run a month would be: 360 gallons of alcohol times 16 gallons of mash for a total capacity of 5760 gallons.

In practice, the farmer would probably use six 1,000 gallon fermenting tanks to supply a 500 gallon still. The beer would be

Table 8-1. Fuel Production Based on Fermenter Tank Size.

Fermenter Size (tanks filled to 80% capacity to allow for foaming). Approximately one gallon of alcohol fuel for about 16 gallons of corn mash.

		55 gal.	100 ga.	250 gal.	500 ga.	1000 gal.
Yearly fuel yield by frequency of use.	fuel yield per run	2.75 gal.	5 gal.	12.5 gal.	25 gal.	50 gal.
	12 times a year (once a month)	33 gal.	60 gal.	150 gal.	300 gal.	600 gal.
	24 times a year (twice a month)	66 gal.	120 gal.	300 gal.	600 gal.	1200 gal.
	52 times a year (once a week)	143 gal.	260 gal.	650 gal.	1300 gal.	2600 gal.
	100 times a year (continuous production)	275 gal.	500 gal.	1250 gal.	2500 gal.	5000 gal.

pumped from the fermenters on a continuous basis until all were empty. Obviously, this much fuel requires a healthy investment in both time and money.

To make about 800 gallons of fuel per year requires less than 70 gallons per run. Twice a month requires only about 35 gallons per run and once a week means about 15 or 16 gallons of fuel per run. In order to make fuel once a week, fermenting and distilling equipment should have the following capacity: 16 gallons of fuel per run times 16 gallons of corn mash per gallon for fuel produced equals 256 gallons total.

This means the small scale fuel producer could use a single 256 gallon fermenter (plus some room for foaming) or about six 55 gallon drums for fermenting and a single 55 gallon drum for a still.

In order to cut down the time spent making fuel to once a month, increase the mash handling capacity about four times. This is roughly 1024 gallons total capacity per month (actually 1067 for a total of 800 gallons of alcohol per year). About 20, 55-gallon drums are needed for fermenters and a 200 to 300 gallon still is needed for once a month fuel production. An alternative method is to use several 500 gallon tanks for fermenting. The choice depends on the cost of several small tanks versus one or two large tanks.

If all this seems confusing, just remember that the capacity of the fermenting and distilling equipment times the frequency of use equals the total amount of fuel produced.

Too small a production capacity means spending all your time brewing and distilling alcohol. Too large a production capacity means needless expense in building or buying fermenting and distilling equipment. Too large a still also means an excess of alcohol over and above your yearly fuel requirements. And according to the law, you can't sell it or even give it away unless you apply for the more complex operators permit. Under the terms of your experimental permit, you are permitted to make fuel for your own use only.

If you do opt for the more complex ATF permit, you can sell your excess fuel to your neighbors, friends, possibly a local service station (to make gasohol) or to other alcohol fuel markets that are bound to develop.

To summarize size requirements, the only advantage of having a large capacity still is that it takes fewer runs to produce the fuel. A smaller still requires more effort and time for the same amount of fuel produced. Ideally, you should size your still according to your fuel needs and free time.

FUEL COSTS

This is the most important—and most difficult—question to answer. In the simplest of terms, it all depends on how costs are computed. In other words, if you base the fuel costs on the prevailing market price of the particular raw material used to make alcohol, the per gallon cost of the furnished product might appear to be rather high.

On the other hand, if you figure the fuel cost on what it actually costs you to produce the feedstock, then the cost per gallon might be surprisingly low.

Farmers have the advantage of being able to grow their own "energy crops" for the cost of the seed, fertilizer, tractor fuel, mortgage and equipment payments. Setting aside only 5 percent of the available crop land for fuel production means those few acres will supply enough energy for the entire farm—greatly reducing the real energy costs of crop production.

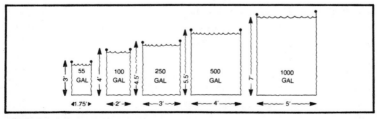

Fig. 8-1. Comparitive tank sizes.

114

City dwellers must either buy crops at the prevailing price from farmers or make use of the resources close at hand. This includes cellulose materials such as scrap paper, wood pulp and garbage, and organic byproducts of manufacturing such as citrus fruit wastes and whey. Often these materials are free for the asking. Fuel costs will vary greatly depending on the source and availability of the raw material.

How To Figure Costs

To estimate the per gallon cost of alcohol fuel, first estimate the cost of the raw material according to what you have to do to produce or obtain it. For example, you might find that corn that sells on the market for $2.00 a bushel but costs 20 cents a bushel to produce. Include the cost of the seed, fertilizer and other chemicals, equipment overhead, mortgage payments and taxes on the land, and other real expenses other than fuel related to crop production, harvesting and transportation. Fuel should be left out of the equation because 5 percent of the crop will be used to supply energy for production.

After you have determined a cost per bushel or per pound, divide that number by the average yield in gallons of alcohol per bushel or pound for that particular crop (see Table 3-1). Corn, for example, yields about 2.5 gallons of ethanol per bushel. This will give you a rough estimate of the cost per gallon fo fuel.

Since most do-it-yourself fuel will be in the 180 proof range (10 percent water left in), deduct 10 percent from the rough cost per gallon. For example, if your rough cost for pure ethanol is approximately 70 cents a gallon, 180 proof fuel would be about 63 cents a gallon.

Deduct the feed value of the stillage. There would be none for most cellulose and sugar feedstocks, but corn and wheat would have a value of about one-third the original material.

Add a per gallon cost for production. This would include what you spent for water to make the mash, malt or enzymes, yeast and heat for cooking and distillation.

Add in an appropriate per gallon cost for the equipment. For example, if you have invested $3,000 in your still and amortize it over five years, that's about $600 per year. If you produce about 1,200 gallons per year, that comes to about 50 cents a gallon—minus your tax depreciation or energy credits.

As you can see, figuring out how much the fuel actually costs you per gallon can get rather complicated. You might find that you can produce alcohol fuel that costs 15 cents a gallon, or $1.50 a gallon. It all depends on how you figure your feedstock costs, stillage value, production costs and equipment investment.

How To Reduce The Costs

Whenever possible, grow your own feedstock or salvage it from another source.

Where possible, feed the stillage to livestock in order to reclaim the feed value of the protein, minerals and unused sugars. This applies primarily to corn and other grains.

Keep your production costs to a bare minimum. Making mash takes a lot of water so pump it from a well, a nearby river, a lake or a stream. Rainwater is relatively pure and easy to collect.

Enzymes are expensive to buy so use malt or grow your own malt. The same goes for yeast. Most breweries maintain their own supplies of yeast in culture mediums to keep costs down.

As for heat, use the cheapest source available; scrap wood, corn cobs, stalks, trash, solar (with a vacuum still or solar still).

Keep your equipment costs down by building a still that is big enough to fulfill your needs but doesn't exceed your needs. Use the least expensive construction materials wherever you can. Why use stainless steel when coated steel works just as well?

Take advantage of every tax credit or other benefit offered by the federal, state and local governments that will reduce your fuel production costs. These include low interest loans, tax breaks and depreciation schedules.

A Cost Alternative

Whether your do-it-yourself fuel costs 15 cents a gallon, 50 cents a gallon, $1.00 a gallon or more, the thing to remember is that having a backyard or on-farm alcohol production facility will enable you to continue despite any Arab oil embargos, Big Oil shortages or government rationing. Under such circumstances, gasoline or diesel fuel would be unavailable at any cost so the ability to make your own fuel would be invaluable.

Switching over to 100 percent do-it-yourself fuel might not be for everyone. In some cases, just having a standby capacity to make alcohol suffices as insurance against any potential cutoff in supply.

Another alternative is to have one tractor or vehicle that runs on alcohol and another that runs on diesel or gasoline. The work load can be split between the two, depending on whichever fuel happens to be cheapest or available at the time.

This dual fuel philosophy won't reduce our dependence on foreign oil or help the market for agricultural products, but it might be more practical for some than the total energy self-sufficiency route. An added advantage is that you only make alcohol when you need it. This cuts down on the time spent making fuel as well as the necessary size of your fermenting and distilling equipment. If you only need enough fuel to meet part of your energy needs, then the capacity of your still can be much smaller.

Financing

Building a still is making a financial investment in your energy future. Because it is a capital investment like any other piece of equipment, it too might be subject to investment tax credits, special energy investments credits and depreciation, depending on the prevailing tax legislation at the time.

The best way to determine what benefits you are eligible for is to sit down with your tax consultant, accountant or a pile of the latest IRS informational booklets. Once you've discovered the exact tax savings to which you're entitled, figure the benefits into your cost figures to help lower your per gallon alcohol costs. Financing your venture might or might not be a problem, depending to the extent of your commitment to alcohol production and how much fuel you really need. If you are only interested in experimenting with small batches to see whether or not you can make alcohol, then financing should be no problem. The costs involved in building a simple still are minimal. Most of the hardware can be scrounged from a junk pile or made from other things.

On the other hand, if you are serious about building an alcohol plant with sufficient capacity to supply a significant portion of your energy needs, then financing becomes a major consideration. Depending on the size and construction materials, the costs can range from a few hundred dollars up to thousands of dollars. A lot will depend on how much of the work you do yourself, whether or not you fabricate your own small-scale plant or buy one from a manufacturer. Materials such as copper and stainless steel are much more costly than plastic and coated steel. Obviously, the more sophisticated the distilling apparatus, the higher will be your investment.

If you are interested in getting into larger scale alcohol production on a cooperative basis with friends or neighbors, then you might be going beyond the realm of do-it-yourself fuel production and getting into the arena of commercial operation. This could be the better route for some. By pooling your resources, you can build an alcohol plant with a capacity of 1 to 3 million gallons a year and launch a money making venture.

The first place you should look for money is in your own savings account. Unfortunately, do-it-yourself financing for most of us is limited to investments under a few thousand dollars. Beyond that, you'll probably need outside help. The second best source of funds should be friends, relatives, neighbors and rich uncles. The Bureau of Alcohol, Tobacco & Firearms allows partnership applications for experimental permits. This way you and your partner(s) can share the fuel as well as the expenses.

For outside financing, you should seek out your friendly banker. His interest rates will be less than those of your friendly finance company—although the bank loan may be somewhat more difficult to obtain. Most bankers are overly cautious by nature, and highly suspicious of anything with the word experimental attached to it. The more risky the proposed venture, the more convincing it will take to get the loan.

Your banker will have a list of "what if's" a mile long. "What if the still doesn't work? What if the thing blows up? What if the price of gasoline drops back down? What if it turns out you don't want to spend the time to make alcohol? What if something happens to you?" They're the same type of questions they ask everyone who wants to carry some of the bank's money out the front door.

Assuming you have a good credit reputation with your bank, you should be able to convince your banker that the still is worth the risk because:

- Being self-sufficient on alcohol fuel will permit you to keep farming (or conducting your business) while others are suffering from shortages of gasoline or diesel fuel.
- You are personally and financially committed to alcohol fuel and that you are serious and sincere about becoming energy independent.

● You have already successfully made alcohol on a small experimental scale and know that it is both practical and economical. You might invite your banker to see your still and a demonstration of the fuel you've produced.

Your toughest job will be to sell the banker on the notion that alcohol fuel production is both possible and practical. It's still a very new field so be prepared to meet with a high degree of skepticism.

Your chances of convincing your banker can be improved if you opt for one of the already proven still designs available from the various manufacturers. This would help eliminate doubts your banker might have about your mechanical ingenuity or ability.

LIST OF MATERIALS

A list of most of the materials and supplies needed to make alcohol is provided below. The list does not include such things as plumbing fittings, pumps or whatever other hardware you might build into your system since the requirements will vary from one still design to the next. For example, a vacuum still requires some kind of vacuum pump. A solar still requires some kind of solar collector, and so on. Notice too, that some of the equipment is listed more than once under several different headings. The same thermometer can be used to check the mash, the fermenter and the still, but it might be more convenient to have several. The same vessel can be used for mash preparation, cooking, fermentation and distillation in a simple still, yet be several different vessels in a larger version.

Mash Preparation

Raw material; sugar, starch or cellulose.
Water; to mix with raw material to make mash.
Malt or enzymes; to convert starches or cellulose to sugar.
Iodine; to test mash for complete starch conversion.
A meter or pH test paper; to test the pH of mash.
Acid and alkaline; to adjust the pH of mash.
Cooking vessel; to prepare and cook mash.
Thermometer; to monitor the temperature of the mash.
Hydrometer; to test sugar concentration.
Heat source; to cook the mash.

Fermentation

Fermentation tank; which can be the same as the cooking vessel.

Yeast; to convert sugar to alcohol.

Fermentation lock; to prevent contamination.

Thermometer; to check the fermentation temperature.

Distillation

Still; to separate alcohol vapor from beer.

Heat source; to heat the still.

Thermometer; to monitor the temperature of the beer in the still.

Safety valve; so that the still doesn't explode.

Distilling column; to separate alcohol and water vapors.

Condenser; to condense alcohol vapors.

Hydrometer; to test the proof of alcohol fuel.

An alcohol fuel storage tank.

Denaturant.

Miscellaneous

An ATF experimental or operating permit; required by law.

A storage or drying tank for stillage.

Cleaning chemicals; to clean tanks after each run.

Chapter 9
A Do-It-Yourself Still

Although building a still is basically a matter of incorporating the distillation principles already covered in this book into a design of your own, it helps to see what others have done to give you some idea of where to begin. The wood-burning still described in this chapter was developed by the people at the *Mother Earth News*, a publication devoted to self-sufficiency, alternative forms of energy and doing more with less.

This particular still will enable you to make 170 proof alcohol at a rate of three-fourths of a gallon per hour—sufficient to meet fuel requirements of about 100 gallons a month or less. Obviously, this isn't enough to make a farm energy self-sufficient but it might be enough to power your pickup truck or car or to meet a portion of your energy needs.

MOTHER'S STILL

This still (Figs. 9-1 and 9-2) can be assembled from readily available or recycled materials, making it a fairly inexpensive unit to build (Table 9-1). It is also inexpensive to operate because it uses wood for the heat source. The basic components of the still are two identical discarded electric water heater tanks—available at a local landfill or from an appliance store junk pile—a few sections of 3-inch copper pipe, some assorted metal stock and some odds and ends of plumbing hardware. The tanks should be of the same diameter and of nongal-

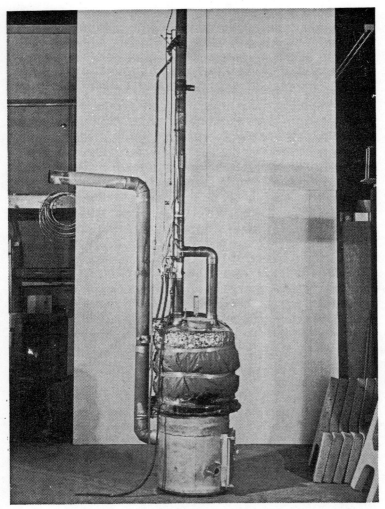

Fig. 9-1. This three-quarter gallon per hour still is made from recycled electric water heater tanks and uses a marble packed distillation column.

vanized metal. Try to get the short, squat tanks that hold 30 to 50 gallons.

The mash solution is contained in a hot water tank and is constantly simmered by a blaze in the firebox underneath. The vapors that rise from the steaming liquid travel up through a short length of 3-inch conduit and into a 5-foot section of pipe (the column) which is packed with glass marbles. A 5-inch

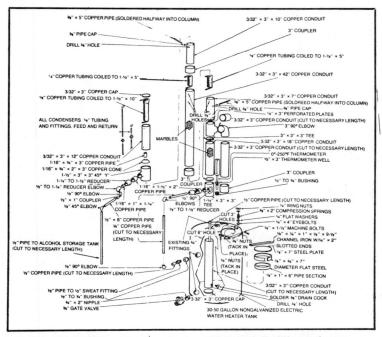

Fig. 9-2. Construction details of a beer tank and distilling column.

length of coiled copper tubing is located at each end of the column.

As the alcohol vapors and steam pass through the marble maze, most of the water will condense onto the cool glass surfaces. The marbles provide a large surface area for this to happen. By running cool water through the coiled tubes at a controlled rate, the condensation process can be regulated to provide the most efficient separation of alcohol vapors and water.

Ideally, only pure alcohol vapor should come out of the top of the column as the water condenses and trickles to the bottom. In practice, however, a small amount of water will make it past the marbles in the form of steam. That is why the finished product is 170 proof and not 200 proof. At the same time, a small portion of the alcohol will also condense and be carried to the bottom of the column with the waste water. Therefore, an additional plumbing circuit is incorporated into the base of the column to route some of this recoverable water/alcohol mixture back into the mash for recycling.

Table 9-1. Bill of Materials for a Still.

Item	Price
(2) 30-50 gallon nongalvanized electric water heater tanks	$ 4.00
14' 3/32" x 3" copper conduit	82.00
5' 1/8" x 2" x 2" angle iron	2.00
7' 1/16" x 1" x 1 1/2" channel iron	3.00
(1) 3/16" x 13 1/2" x 13 1/2" steel plate	1.00
4' 3/16" x 1" flat steel	1.00
10' 3/16" x 3" flat steel	3.50
40' 3/8" concrete reinforcing bar (rebar)	4.40
(1) 1/6" x 8" x 22" flat plate	2.00
(1) 3" x 3" Schedule 40 pipe	.35
(1) 4" x 6" Schedule 40 pipe	.75
2' 1/8" x 1 1/2" x 1 1/2" angle iron	—
(2) 4" hose clamps	.95
(2,500) glass marbles	17.34
100' 1/8" copper tubing (1/4" O.D.)	43.29
10' 3/8" (O.D.) copper pipe	.25
(2) 3/8" copper pipe caps	.20
(3) 3" couplers	10.86
(1) 1 1/2" x 3" x 3" tee	3.98
(1) 3" x 3" x 3" tee	4.18
(1) 1 1/2" x 3" x 3" x 3" 45° "Y"	4.09
(1) 3" 90° elbow	3.23
10' 1/2" rigid copper pipe	4.56
(4) 1/2" copper 90° elbows	1.48
(2) 1/2" copper 45° elbow	.35
(1) 1/16" x 1 1/2" x 4" copper pipe	.70
(1) 1 1/4" to 1 1/2" reducer	.75
(1) 1/2" to 1 1/2" reducer	.80
(1) 1/16" x 1" x 1 1/4" copper pipe	.30
(1) 1/2" x 1" coupler	.45
(1) 1/2" to 1 1/4" reducer elbow	.45
(1) 1/2" sweat to 1/2" pipe fitting	.30
(2) 1/2" to 3/4" bushing	.35
(1) 1/2" x 3" thermometer well	1.05
(2) 0°F-212°F straight thermometers	8.25
(1) 0°F-250°F threaded thermometer	6.48
(1) 3/4" x 2" nipple	.55
(1) 3/4" gate shutoff valve	3.10
(1) 3/8" drain cock	.95
(1) 1/16" x 3/4" x 3" copper pipe	.50
(1) 1/16" x 6" x 10" copper plate	1.65
(2) 1/4" x 4" eyebolts	.60
(4) 1/4" flat washers	.10
(2) 3/8" x 2" compression springs	.60
(2) 1/4" wing nuts	.10
(6) 3/8" hex nuts	.12
(3) 3/8" x 1" machine bolts	.15
(1) 3/8" x 2" machine bolt	.05
(2) 1/4" x 1 1/2" machine bolts w/nuts	.15
(1) 3/8" flat washer	.20
(1) 3/8" x 1" x 1" compression spring	.02
(1) 1/8" x 1" x 6" pipe section	.25
(2) 1/8" tubing threaded unions (if desired)	.30
(6) 1/8" sweat to 1/4" pipe fittings	.90
(6) 1/4" pipe to 1/4" hose barbs	1.05
(2) 1/4" brass tees	.65
(2) 1/4" male to female brass elbows	.70
(3) 1/4" needle valves	6.42
(1) 1/2" brass tee	.45
(2) 1/4" to 1/2" bushings	1.00
(2) 1/4" close nipples	.30
(1) 1/2" sweat to 1/2" pipe fitting	.24
(2) garden hose to 1/2" pipe adapters	.85
(1) garden hose "Y" adapter	.98
12' baling wire	—
3' 1/4" air hose (cut to necessary length)	1.48
(1) 4" stovepipe 90° elbow	2.09
(1) length 4" stovepipe	—
1 piece fiberglass insulation batting (if desired)	—
Total Cost	$244.42

Note: The figures above represent new material prices. By scrounging and buying from salvage or scrap dealers (especially with regard to the copper items), the total cash outlay for your wood-burning still project can easily be halved.

At the top of the column just above the upper cooling coil is an inverted funnel which is surrounded by more cooling coils. When the "stripped" alcohol vapors pass through this cone and make contact with the coils, the ethanol liquifies, runs past the outer surface of the funnel, flows through a downspout and enters the storage container.

THE FIREBOX

Before you start still building, you will need an acetylene torch (with welding and cutting tips), a pipe cutting tool, several C-clamps or vicegrips, a hammer, a ruler and a file.

Begin by cutting around the circumference of one of the salvaged water tanks at a height of about 18 inches (Fig. 9-3). Fashion an opening in the side of the tank to accommodate the firebox door. Next, weld together an angle iron frame for this opening and fasten it to the tank. Cut a steel plate to serve as the portal for the cooker. Border the inner surface of this lid by welding four sections of trimmed-to-size channel iron in place, then go on to fabricate two hinges and the latch system.

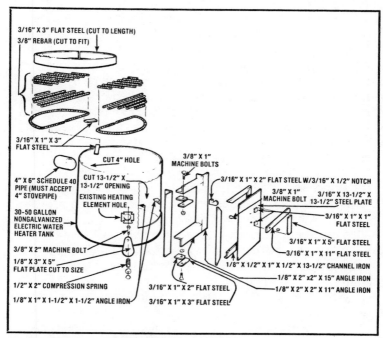

Fig. 9-3. Construction details for a still firebox.

With these steps done, weld a bolt (head first) to your tank at a point just above an existing heater element hole. Make a small "draft" door from a piece of steel and attach it to this bolt. Use a spring, washer and nut so that the plate can pivot. Cut a hole in the rear of the new firebox and weld the 4-inch stovepipe collar to the circumference of this opening. Then fabricate a grate from three-eights inch rebar and tack the supporting tabs in place.

Weld the three-sixteenth inch by 3-inch to the outside of the water container to serve as a girth band. This midriff support is actually the only joint between the firebox and your mash vat. Therefore, it must be securely fastened to both containers. You can assure a tight seal by welding one end of the iron strap to the fire chamber and guiding the band around the circumference of the vessel with your torch and hammer, tack welding as you go. When the circle is completed, force the remaining water tank into the iron belt border atop the firebox section and run a continuous bead along both upper and lower edges of the metal band.

BUILDING THE STILL

With the firebox and mash tank complete, you are ready to start construction of the still. First, cut two holes in the top of the tank. The holes must be large enough to fit the 3-inch brass tee and the coupler. Then cut a third access hole into the dome of the container. Fashion a sealable lid for this opening by welding a collar around the hole, attaching a slightly larger ring to a piece of circular flat plate, and making a "dog and bar" latch. The edge of this cap can also be lined with silicone sealant to prevent any vapor loss during the distillation process. Braze the tee fitting and coupler to the tank to assure a solid joint.

For the next step, cut one length each of 3-inch conduit and one-half inch conduit, both long enough to reach nearly from the bottom of the tee to the base of the container. Then seal the lower end of the larger pipe with a cap and install a drain cock. Solder this entire assembly to the lower collar of the tee. Take the length of one-half inch conduit and add a 90 degree elbow to the top so that a return line can be formed through the tee atop the tank.

You can now install a three-fourth inch gate valve in the base of your container and a thermometer at its top. Most tanks will already have threaded fittings in convenient locations, so it's just a matter of installing adapters and attaching the hardware.

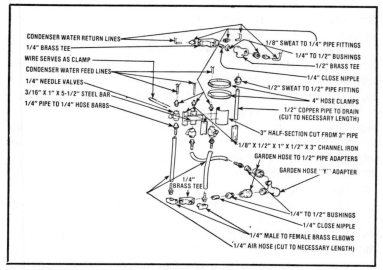

Fig. 9-4. Construction details of water cooling lines.

From this point, you can complete the remainder of the column as follows: Trim each section of three-inch conduit to the lengths indicated in Fig. 9-4. Then make three perforated discs cut to fit snugly inside the tee fittings. They should rest agains the lip inside the three-way unions. Drill each with one-fourth inch holes about one-eighth of an inch apart. Place one of your discs into the top opening of the tee and solder the 18-inch length of pipe in position.

Next, fill this section of conduit with glass marbles, place the second perforated disc into the 3-inch × 3-inch × 3-inch tee and solder that fitting into place. At this point you can permanently fasten the pipe connection between the coupler brazed to the top of the mash container and the horizontal arm on the brass tee.

Install the final drilled plate at the top of the tee and stack the remaining components in the order shown. Don't forget to install the thermometer wells in the two sections of conduit as illustrated. These cylinders must protrude into the pipe, but must be capped on their inside ends to prevent vapor leakage. Be sure to fill the long section of 3-inch conduit with marbles before sealing it up.

The condenser coils can be made by wrapping the copper tubing around a piece of 1½-inch pipe until you produce a coil of

127

the desired length. Be sure to solder seal the joints between the column and the protruding copper piplets.

Unless you have a small funnel that fits snugly in the 45-degree Y at the top of tower you will have to make one. Be certain the funnel is leak free and carefully soldered to the Y or some of the distilled alcohol might leak back into the mash vat.

Complete the copper plumbing circuit to your alcohol storage tank and hook up the three pairs of water feed and discharge lines as indicated. Be certain that each feed line is equipped with a needle valve, since careful water control is essential to the distillation process.

THE FIRST RUN

First, wrap the boiler tank with fiberglass insulation. Next, fill the tank with the fermented beer to a level several inches below the lip of the access hole. This might total 25 gallons or more, depending on how large the water tank is. Next, connect the stovepipe to an outside chimney (if your still is indoors), build a fire in the firebox and wait for the mash to generate some alcohol/water steam. This will occur when the tank thermometer reads about 170 degrees F. Remember, this thermometer should be placed to read vapor temperature, not liquid temperature.

Open the three condenser flow valves and wait for the vapors to reach the 175 degree F to 190 degree F range. At this point, the marbles will begin to crackle—and you can begin fine tuning the still controls. The main condenser coil at the top of the column doesn't need continual adjustment, just let the water flow through it at a slow, steady rate. The other two coils, however, must be controlled very carefully.

During the first two hours or so of distillation, the temperature at the lower primary coil should be maintained at a point between 176 degrees F and 182 degrees F, while the heat in the vicinity of the upper condenser should be held to about 170 degrees F. To cool either area, just increase the flow of water through the coiling coils.

When everything is adjusted properly, an 85 percent pure alcohol product will begin to flow into your storage container. The stream should be slow but steady. If it's excessive, chances are that there is too much water in the alcohol. This undesirable situation can be prevented by cooling the upper and lower coils slightly—but do try to stay within the temperature ranges suggested.

Be certain to check the fire regularly. During most of the run, the vapor temperatures in the mash tank will stay around 180 degrees F. But as more alcohol is driven from the solution, that figure will rise. When the heat in the vat reaches approximately 200 degrees F, after perhaps three hours, the condensed product will be mostly water. Shut the apparatus down to prevent dilluting the alcohol with water.

It will take a few runs to get the feel of how to control the still for best results. The tail end of the run will be a fairly weak proof but the first two-thirds of the run should yield fuel in the range of 170 proof. This still can produce about three gallons of fuel from about 30 gallons of beer.

To increase the proof of the finished product or to remove excess water that might have gotten through, you can run the distilled product through the still a second time. This procedure will drive most of the excess water out of the solution and yield a nearly pure fuel of 190 proof (95 percent alcohol). You might also want to save the "tail end" of each run, which will be about 60 to 100 proof to recycle it with the beer for the next run. This will increase the overall yield on the next batch you run through the still. Fully illustrated and detailed construction plans for this still are available for $15.00 by writing to:

The Mother Earth News
Dept. W.
P.O. Box A
East Flat Rock, NC 28726.

Chapter 10
Ready-Made
Stills and Supplies

For those who lack the time or mechanical aptitude to build a still, or would prefer to go with a "proven" design rather than a concoction of their own engineering, there are a number of firms who are offering ready-made stills. These range in size from tabletop models, sufficient to make enough fuel to power a lawnmower or moped (Fig. 10-1) to plants capable of producing 20 to 30 gallons of alcohol per hour on a continuous basis.

Easy Engineering offers a plant for the large farm. It will produce 200 to 250 gallons of ethanol per day. The plant includes a 1200 gallon batching tank, three 1200 gallon fermentation tanks, steam coils, an agitator, control panel, steam generator, two distilling columns, a 250 gallon denaturing tank, a 1000 gallon holding tank and all pumps and hardware. The price for this complete plant is around $50,000 F.O.B. Denver, CO.

Energy Restoration Inc. offers a plant that will produce 30 gallons of 190 proof fuel per hour. The kit includes all tanks, columns, automatic controls and hardware and sells for $42,000.

Fig. 10-1. Tabletop still capable of producing enough fuel for a lawnmower or moped.

At the other end of the spectrum, *Jennings Enterprises* offers a tabletop still that consists of a 12-quart boiler, separator and condensing coil. Although it doesn't produce much fuel, it does make a good still to experiment with.

STILL MANUFACTURERS

For current information regarding prices, catalogs and equipment, write to the following:

ACR Process Corp.
602 East Green St.
Champaign, IL 61820
(217) 384-8003

Agri Stills of America
3550 Great Northern Ave.
Springfield, IL 62707
large scale alcohol plants

Alternative Energy Ltd.
Route 1
650 Pine St.
Colby, KS 67701
(913) 462-7171

Davy McKee Corp.
10 S. Riverside Plaza
Riverside, IL 60606
(312) 454-3685
large scale engineering firm

Easy Engineering
3353 Larimer St.
Denver, CO 80205
(303) 893-8936

Energy Restoration, Inc.
Suite 403, Century House
1201 J. St.
Lincoln, NE 68508
(402) 435-1379

Glitsch, Inc.
P.O. Box 226227
Dallas, TX 75266
(214) 631-3841

Jennings Enterprises
P.O Box 9570
Panama City Beach, FL 32407

R.B. Industries
P.O. Box 82
Riverdale, MI 48877
(517) 833-7584

SUPPLY SOURCES

The object of do-it-yourself fuel production is to do as much of it yourself as possible. But not everyone can grow yeast or malt. And some things such as pH test paper, hydrometers and thermometers have to be purchased. As for the fermentation tanks, cookers and stills, you should first try to build them yourself or make them from some other type of container such as an old drum water tank.

A list of suppliers has been included to help solve the problem of finding hard-to-get supplies and equipment. Before calling or writing these firms, you should first check the Yellow Pages of your local phone directory under the appropriate headings. If you live in a rural area or in a small community, the local library will have phone directories for nearby large cities. Check under the following headings for:

Yeast
Yeast ....Bakery Supplies

Enzymes
Enzymes ....Brewery Supplies

Laboratory Equipment
Hydrometers,   Scientific Apparatus & Instruments
pH paper and   Thermometers
thermometers ....Brewery Supplies

Malt
Brewery Supplies
Malt...Brewers

Vermentation tanks,   Tanks & Vats
cooking vessels ....Brewery Supplies

Stills ....Distilling Apparatus

The list is by no means complete but does include most of the major suppliers.

Acme Process Equipment Co.
Oreland Mill Road
Oreland, PA 19075
(215) 886-8600
Vats & Tanks

Advanced Instruments Inc.
1000 Highland Ave.
Needham Heights, MA 02194
(617) 449-3000
hydrometer & test equipment

American Malting Inc.
100 Childs St.
Buffalo, NY 14203
(716) 852-3648
Malt

Anheuser-Busch, Inc.
Industrial Products Division
10877 Watson Road
St. Louis, MO 63127
(800) 325-3917
yeast

A.P.V Co. Inc.
395 Fillmore Ave.
Tonawanda, NY 14150
(716) 692-3000
distillation equipment

Bailey Controls Co.
29801 Euclid Ave.
Wickliffe, OH 44092
(216) 943-5500
control equipment

BASF Wyanditte Corp.
Chemical Specialities Division
Wyandotte, MI 48192
(313) 382-6000
cleaning chemicals

Bauer-Schweitzer Malting Co. Inc.
530 Chestnut St.
San Francisco, CA 94133
(415) 362-1617
malt

Bishopric Products Co.
4413 Kings Run Drive
Cincinnati, OH 45232
(513) 641-0500
vats & tanks

Buffalo Malting Co.
375 Grain Exchange Building
Minneapolis, MN 55415
(612) 339-9171
malt

Chemtech Industries Inc.
9909 Clayton Road
St. Louis, MO 63124
(314) 997-4600
cleaning chemicals

Chilton Malting Co.
137 E. Main St.
Chilton, WI 53014
(414) 849-2338
malt

E.M. Adler
120-32 83rd Ave.
Kew Gardens, NY 11415
(212) 846-4696
malt

Enzyme Development Corp.
2 Penn Plaza
New York, NY 10001
(212) 736-1580
enzymes

F. Bing Inc.
1860 Broadway Ave.
New York, NY 10023
(212) 581-6090
malt

Fleischmann Malting Co. Inc.
410 Grain Exchange Building
Minneapolis, MN 55415
(612) 338-4771
malt

Fischer and Porter Co.
108 Warminister Road
Warminster, PA 18974
(215) 674-6000
Hydrometer & Test equipment

Gamlen Chemical Co., Sybron Corp.
333 Victory Ave.
San Francisco, CA 94080
(415) 873-1750
cleaning chemicals

G.B. Fermentation Industries, Inc.
One N. Broadway
Des Plaines, IL 60016
(312) 827-9700
Enzymes, yeast

Johnson & Carlson
848 Eastman St.
Chicago, IL 60622
(312) 664-1580
vats & tanks

Joseph Oat Corp.
2500 Broadway
Camden, NJ 08104
(609) 541-2900
vats & tanks

Micro Tec Laboratories
Rt. 2 Box 19
Logan, IA 51546
(712) 644-2193
Hydrometer & test equipment

Miles Laboratories
Marschall Division
1127 Myrtle St.
Elkhart, IN 46514
(219) 264-8111
Enzymes

Mother Earth News
Dept. W.
P.O. Box A
East Flat Rock, NC 28726
(800) 438-0238
*enzymes, hydrometer
& testing equipment*

N L Industries Inc.
Industrial Chemicals Division
P.O. Box 700
Highstown, NJ 08520
(609) 443-2000
Enzymes

Nooter Corp.
P.O. Box 451
St. Louis, MO 63166
(314) 621-6000
vats & tanks

Novo Laboratories Inc.
59 Danbury Road
Wilton, CT 06897
(203) 762-2401
Enzymes

RABCO Scientific
8935 J. St.
Omaha, NE 68127
(402) 331-1233
Hydrometer & Test equipment

Rascher & Betzold Inc.
5410 N. Damen Ave.
Chicago, IL 60625
(312) 275-7300
Hydrometer & Test equipment

Rohm and Hass Co.
Independence Mall West
Philadelphia, PA 19105
(215) 592-3000
Enzymes

**Schwartz Service
International Ltd.**
230 Washington St.
Mt. Vernon, NY 10551
(914) 664-1100
Enzymes

Siebel Sons Co.
Enzyme Product Div.
4055 W. Peterson Ave.
Chicago, IL 60646
(312) 463-3400
Enzymes

Standard Brands
Fleischmann Bakery Division
Charles Point
Peekskill, NY 10566
(914) 737-2900
Yeast

Universal Foods Corp.
433 East Michigan Ave.
Milwaukee, WI 53201
(414) 271-6755
Yeast

**U.S. Army Pollution Abatement
Division**
Natick Development Center
Natick, MA 01760
(617) 653-1000
Enzymes

Wilkens-Anderson Co.
4525 W. Division St.
Chicago, IL 60651
(312) 384-4433
Hydrometer & test equipment

Chapter 11
Rules & Regulations

As with any enterprise undertaken these days, there is the unavoidable government red tape to reckon with. In short, to make alcohol legally you must:

—Obtain a permit from the U.S. Bureau of Alcohol, Tobacco & Firearms.
—Post a surety bond.
—Comply with any other state or local regulations concerning the production of ethyl alcohol.

You could build a secret still and become a bootleg fuel producer or you could try pleading ignorance of the law before the judge. But unless you want to risk the consequences of possible civil or criminal charges, it's wisest to play by the rules of the game. Despite rumors, the paperwork is neither difficult nor particularly time-consuming for a federal "experimental" permit on a small scale fuel alcohol production still. A large-scale commercial plant, however, is another story. For that, you will need a good lawyer to decipher all the EPA, OSHA and ATF forms and regulations.

FEDERAL PERMIT

Because Uncle Sam derives revenue from taxes on drinking alcohol, the production and distribution of all ethyl alcohol, regardless of use, comes under the watchful eye of the U.S.

Bureau of Alcohol, Tobacco & Firearms. The two types of permits are an operating permit and an experimental permit.

An operating permit authorizes alcohol to be produced for sale or any other nonexperimental purpose. This is the type of permit needed to operate a large-scale commercial plant and it involves a lot of paperwork, bookkeeping and supervision. For instance, a government inspector must be present to supervise the denaturing of the alcohol fuel and special security precautions must be observed. This means periodic visits by the local ATF field inspector to make sure of compliance with the letter of the law.

The experimental permit authorizes you to "establish an experimental distilled spirits plant (a still for experimentation in, or development of: Sources of material from which spirits may be produced; Processes by which spirits may be produced or refined; or Industrial uses of spirits."

An experimental permit gives you permission to build and operate an alcohol fuel still. You can make all the fuel you want for your own use—but you can't sell the stuff or even give it away. For that, you must have an operating permit.

An experimental permit is valid for two years. At the end of that time, it will expire unless you reapply.

EXPERIMENTAL PERMIT

The home production of alcohol for use as an alternate fuel is a rather new concept to the ATF. You must write to the ATF to request their permission to make alcohol before you can begin production. Since there are no forms to fill out for an experimental permit, submit a letter describing your intentions and wait for the reply. Depending on the workload, you will receive an answer in about two to three weeks.

Your letter to the ATF should contain the following information:

A General Statement. For example, you might tell them you are interested in experimenting with solar energy to make alcohol fuel for a tractor, combine, truck, car or whatever. Your statement need not sound like a Philadelphia lawyer wrote it. Just stick to the basics and be sure to clearly state that your intent is to make alcohol for fuel use only.

The Plant Location. If you are a farmer, the description should include your entire farm. This is because an experimental permit does not allow you to remove alcohol from the premises unless it is being used to fuel your car or truck. Include in the description the number of acres involved. You should also describe the buildings that will be used in the production and storage of the fuel alcohol and their relative location on the farm.

The Production Process and Still Equipment. Describe what raw materials you will be using to make mash, such as corn, spoiled grain or kitchen garbage, and a fairly detailed description of how you plan to distill the fuel. For example, "I plan to recover ethyl alcohol from the mash in a still of my own design. It will consist of a large vat, a firebox and steam boiler, two distilling columns, a condenser and alcohol fuel storage tank." Be sure to include all equipment from the mash tank to the storage tank. In addition, describe when and how you intend to denature the alcohol produced. The ATF prefers you denature the alcohol immediately after its manufacture by mixing in gasoline or kerosene in a 1:10 ratio (one gallon of gas to ten gallons of alcohol). This should be done in the fuel storage tank and will discourage anyone from drinking your home brew fuel.

Security Precautions. The ATF wants to be sure no one is going to steal the fuel alcohol to make moonshine. Be sure to tell them that the shed in which the alcohol is to be stored will be locked, or that the storage tank will be padlocked. If you have a watchdog, mention that too.

The Rate of Production. The ATF wants to know how much alcohol you will be making so that they can set your bond accordingly. You should state in gallons the amount of alcohol you expect to produce in an average 15-day period. Tell them the average proof of the finished product. For example, "I plan to make approximately 100 gallons of alcohol, averaging between 160 to 190 proof, in a 15-day period." Be careful when making your estimates because if your annual production exceeds 2,500 gallons, the amount of your bond will increase dramatically.

Signature, Name, Address and Phone Number. If you have a partner, be sure that you *both* sign the letter.

ADDRESSES OF ATF REGIONAL OFFICES

Your letter of application should be mailed to your regional ATF office for consideration.

Central Region

Indiana, Kentucky, Michigan, Ohio, West Virginia
    Regional Regulatory Administrator
    Bureau of Alcohol, Tobacco
    and Firearms
    550 Main Street
    Cincinnati, OH 45202

Mid-Atlantic Region

Delaware, District of Columbia, Maryland, New Jersey, Pennsylvania, Virginia
    Regional Regulatory Administrator
    Bureau of Alcohol, Tobacco
    And Firearms
    2 Penn Center Plaza, Room 360
    Philadelphia, PA 19102

Midwest Region

Illinois, Iowa, Kansas, Minnesota, Missouri, Nebraska, North Dakota, South Dakota, Wisconsin
    Regional Regulatory Administrator
    Bureau of Alcohol, Tobacco
    and Firearms
    230 S. Dearborn Street
    15th Floor
    Chicago, Illinois 60604

North-Atlantic Region

Connecticut, Maine, Massachusetts, New Hampshire, New York, Rhode Island, Vermont, Puerto Rico, Virgin Islands
    Regional Regulatory Administrator
    Bureau of Alcohol, Tobacco
    and Firearms
    6 World Trade Center, 6th Floor
    (Mail: P.O. Box 15,
    Church Street Station)
    New York, NY 10008

Southeast Region

Alabama, Florida, Georgia, Mississippi, North Carolina, South

Carolina, Tennessee
Regional Regulatory Administration
Bureau of Alcohol, Tobacco
and Firearms
3835 Northeast Expressway
(Mail: P.O. Box 2994)
Atlanta, GA 30301

Southwest Region

Arkansas, Colorado, Louisiana, New Mexico, Oklahoma, Texas, Wyoming
Regional Regulatory Administrator
Bureau of Alcohol, Tobacco
And Firearms
Main Tower, Room 345
1200 Main Street
Dallas, TX 75202

Western Region

Alaska, Arizona, California, Hawaii, Idaho, Montana, Nevada, Oregon, Utah, Washington
Regional Regulatory Administrator
Bureau of Alcohol, Tobacco
and Firearms
525 Market Street
34th Floor
San Francisco, CA 94105

BOND

Once your letter of application has been processed—delays can be caused by forgetting to include all the necessary information—you will be required to post a surety bond with the government.

Even though you pay no taxes on fuel alcohol—only drinking alcohol—you are still required by law to post a bond. This is to cover any excise tax liabilities that would be due should any of your "fuel" alcohol find its way into someone's shot glass. The current rate of taxation on drinking alcohol is $10.50 per 100 proof gallon, or $21 per gallon of pure alcohol.

If you plan to produce less than 2,500 gallons of fuel alcohol annually in your experimental still, you will be required to post a $100 bond.

The bond can be underwritten by securities or in the form of a cashier's check or money order made payable to the Internal Revenue Service (not the ATF). No personal checks will be accepted. The government will hold the money for two years, the duration of your experimental permit, at the end of which you will receive either a refund (without interest or you can renew the permit application.

If you plan to make more than 2,500 gallons of fuel alcohol per year, you must submit a bond in an amount equal to the number of gallons to be produced in a 15-day period multiplied by the tax rate of $21 per gallon. The total must then be rounded up to the next higher hundred.

For example, an estimated production of 150 gallons in a 15-day period would figure out as follows:

150 gallons × $21 tax per gallon = $3,150
round total up to the next higher hundred = $3,200

As you can see, estimating an alcohol production over 2,500 gallons a year carries with it a rather steep bond. By going from an estimated production of 100 gallons in a 15-day period (about 2,400 gallons per year) to 150 gallons, the amount you're required to post jumps from $100 to $3,200.

Needless to say, giving the government $3,200 interest free for two years is great for the national debt but it doesn't do much for your pocketbook. Your best bet is to buy the necessary bond coverage from a bonding company. It's like buying an insurance policy. You pay a certain fee for a given amount of bond coverage. You are out the fee but it might be cheaper than losing the interest on your money for two years—not to mention the loss due to inflation.

The following is a sample statement of deposit to be submitted with surety bond:

I have deposited with the Bureau of Alcohol, Tobacco and Firearms, A Cashier's Check or money order in the amount of $_____, as security for the faithful performance of any and all the conditions or stipulations of a certain obligation entered into by me with the United States, under date of_____, on Form 2601, which is made a part hereof. I agree that, in case of any default in the performance of any of the conditions and stipulations of such undertaking, the Bureau of Alcohol, Tobacco and Firearms shall have full power to apply in whole or in part, the amount on deposit to satisfy any damages, demands, or deficiency arising by reason of such defaults. The undersigned farther agrees that the authority herein granted is irrevocable. I further agree that no interest will be paid to me on the amount deposited.

Signature and Title

FIELD INSPECTION

After you have completed the application and filed the bond, a field inspector from the ATF will drop by to make sure everything is as reported. Note : solar still applications are given advance approval with no inspection required. Upon completion of the inspection, the agent will file a written report with the ATF. You will then be issued a formal authorization to begin operations. You are not supposed to make any alcohol prior to receiving this authorization.

All ATF officers have the right to gain access to the plant premises for purposes of inspection. If you do not own the property on which your still will be located, the legal owner must grant in writing permission for ATF inspectors to inspect your still. No permission means no permit.

The ATF inspector will also give you Forms 4805 and 4871 which deal with water quality considerations and environmental information. These forms are to be completed and returned to the inspector while he is at your premises. The purpose of such forms is to make sure you aren't planning to pollute the local water supply by dumping mash residue in a nearby creek or pollute the air by burning old tires to fire your still.

LOCAL REGULATIONS

The ATF permit does not exempt you from complying with any state or local regulations concerning the production of alcohol. Most rural areas are not encumbered with "bureaucratic supervision" as are most metropolitan areas. Nevertheless, there are some things you should be aware of.

For example, the state fire marshall might have something to say about storing and handling alcohol fuel. Alcohol is usually in the same category as gasoline under flammable liquids, so the same fire protection and safety standards would apply to both. Most regulations require that fuels be stored at a distance from buildings and that explosion-proof wiring, proper valving and emergency venting may be used in any building where alcohol is made or stored. Safety standards might also be

required to meet the qualifications for insurance coverage. Check with your insurance agent on this.

Local air and water quality standards might also have a bearing on your operation. Special permits might be required in some areas to dispose of waste water or other residues.

County and city building codes and zoning ordinances might also put the dampers on your planned alcohol production facility. Before building anything more than a small-scale still, check your local ordinances.

For a more detailed description of the specific ATF regulatory requirements governing alcohol production, write to your regional ATF office and request: *Ethyl, Alcohol for Fuel Use,* ATF P 5000.1 (9-78). This booklet is also available from the Superintendent of Documents, U.S. Government Printing Office, Washington, D.C. 20402. Request stock number 048-012-00045-1.

QUESTIONS AND ANSWERS

The following questions have been taken from an official ATF handout. They are included to further clarify certain points regarding the operation of an experimental "distilled spirits plant" and the application procedure.

Question: Can I sell or loan any excess alcohol produced to another person for fuel use?

Answer: No. The alcohol produced may be used as fuel only at the plant premises described in your bond and letter application. Only a plant qualified as a commercial DSP can sell, loan or give alcohol to another party.

Question: Can I remove some of the alcohol from the plant premises for my own use? (Example: as fuel for my personal car).

Answer: You may remove alcohol from your plant premises for your own use as a fuel; however, the alcohol must be *completely* denatured according to one of the two formulas listed below before removal. Formula No. 18. To every 100 gallons of ethyl alcohol add:

—2.5 gallons of methyl isobutyl keytone;

—0.125 gallon of pyronate or a compound similar thereto;

—0.50 gallon of acetaldol; and

—1 gallon either kerosene or gasoline.

OR

Formula No. 19. To every 100 gallons of ethyl alcohol add:

—4.0 gallons of methyl isobutyl keystone; and

—1.0 gallon of either kerosene or gasoline.

Question: Must I denature the alcohol before using it on my farm?

Answer: While we can approve applications where good cause is shown for the need to use denatured spirits as fuel, we prefer that you denature your alcohol with gasoline, diesel fuel or heating fuel immediately after production. This will allow us to approve less stringent security systems and recordkeeping requirements than what we impose on applicants who do not denature their alcohol.

Question: Can any of the alcohol produced be used for beverage purposes?

Answer: Absolutely not. Besides the IRS Excise Tax of $10.50 per proof gallon, which you would become liable for, you could also incur severe criminal penalties.

Question: Can I build my distillery system prior to receiving any authorization from ATF to operate?

Answer: Yes. However, you must file an application to establish an experimental distilled spirits plant with ATF immediately after its completion. You may, however, file sooner if you feel you will have the equipment set up prior to the qualification visit by the ATF inspector. However, under no circumstances may you start producing spirits prior to receipt of a formal authorization by the Bureau.

Question: Can I qualify two or more farms for my plant premises?

Answer: Yes, if they are in close proximity to each other so as to allow an ATF inspector to inspect all premises without causing undue travel and administrative difficulties.

Question: How long a period is the authorization effective?

Answer: We are currently approving operations for a two year period unless you include in your application some justification for a longer period of time.

Question: Can I renew my experimental plant authorization after expiration?

Answer: Yes. When the authorization expires, you may file a new application listing the current information on all subjects originally described (security, rate of production, equipment, etc.). A new bond will not be required unless significant changes in operations have occurred since the original filing.

Question: Can partnerships and corporations make an application as well as individuals?

Answer: Yes, however, if the application is filed by a partnership, all partners must sign it. If it's filed by a corporation, a person authorized by the corporation must sign and proof of such authorization must accompany the application (e.g. certified copy of a corporate resolution or abstract of bylaws giving such authority).

Question: Can I convert to a commercial operation?

Answer: There is no simple means of converting an experimental operation into a commercial operation. Normally all provisions of Title 27, U.S.C. Chapter 51 and Title 27, CFR Part 201 will be waived for an experimental alcohol fuel-related DSP except for those relating to:

- —Filing application for, and receiving approval to operate an experimental DSP for a limited, specified period of time.
- —Filing of a surety bond to cover the tax on the alcohol produced.
- —Attachment, assessment and collection of tax.
- —Authorities of ATF Officers.
- —Maintenance of records.

No such blanket waiver will be given for commercial operations. You will have to follow the qualification procedure outlined in ATF P 5000.1. We will, however, give favorable consideration to alternate procedures from regulations which do not present a definite jeopardy to the revenue. However, each such variation will be viewed and ruled upon on an individual basis.

Chapter 12
Alcohol Fuels

Alcohol can be used as a motor fuel either straight, blended with other traditional fuels such as gasoline or diesel, or a supplemental fuel in a dual fuel engine.

A QUICK OVERVIEW

The first thing to understand about alcohol fuels is that there is no one fuel that will satisfy every application. Switching a gasoline engine over to run on straight alcohol can be accomplished with a few simple modifications, but not everyone has the ability or motivation to make such changes. Since Detroit stopped making cars that would burn straight alcohol over 70 years ago, the only alternative left to much of the motoring public is to use gasohol (90 percent gasoline and 10 percent ethanol).

For those who wish to convert to straight alcohol, there might be a problem of fuel availability. A farmer or do-it-yourselfer who has the resources and ability to build and operate a still can think seriously about using alcohol in farm equipment or vehicles. But the poor city apartment dweller who is hard pressed even to find a parking space for a car can't make fuel. And finding a source to buy straight alcohol fuel might be a real problem. Another option would be to switch from gasoline to gasohol. Gasohol requires no modifications, is readily available in many areas of the country and is, in many ways, superior to straight gasoline. It has a higher octane rating, a reputation for improving mileage and performance, and sells for about the same price as unleaded gas—depending on the local tax exemptions granted gasohol.

From a nationwide standpoint, gasohol is probably the most practical end use for alcohol in the immediate future. It can help extend our dwindling supplies, of petroleum, reduce our dependence on foreign oil somewhat, and create the incentive and market necessary to expand the emerging alcohol fuel industry. But gasohol is primarily a question of politics rather than technology, so for the do-it-yourselfer it holds little immediate value.

The quickest way to cut the energy strings now held by Big Oil, OPEC and our own bureaucrats is to switch over to straight alcohol. For the do-it-yourselfer with the means and ambition, it holds tremendous promise. Energy self-sufficiency will enable the individual to survive despite shortages, rationing, cutbacks and ever increasing prices.

As alcohol becomes accepted as a viable alternative, it's likely that Detroit and the agricultural equipment manufacturers will see the light and offer alcohol fuel modifications as a factory option to those who request it. One thing is for sure, private enterprisers are rapidly recognizing the opportunities at hand. It should only be a matter of time until alcohol conversion kits become widely available.

STRAIGHT ALCOHOL AS A MOTOR FUEL

Methyl alcohol is quite similar in many respects to ethanol but because the primary method of manufacture is beyond the scope of the do-it-yourselfer (most is made from natural gas for industrial use) it has not been included as part of this book.

Here is a quick summary of straight ethanol alcohol as a motor fuel:

It can be used as a substitute for gasoline in any spark ignition engine—but its use does require some modifications.

The basic changes necessary to use ethanol are to enlarge the carburetor jets about 30 percent in diameter and to advance the ignition timing several degrees. Additional modifications can be made to improve performance and mileage.

Switching to straight alcohol will increase fuel consumption up to 50 percent, depending on the application. However, increasing the compression ratio can narrow this difference significantly.

Switching to straight alcohol can also increase overall horsepower up to 15 percent or more, again depending on the application and type of modifications made.

If straight ethanol is used in a 2-cycle engine such as a motorcyle, snowmobile or chain saw, a vegetable base oil should be mixed with the fuel in place of the regular 2-cycle engine oil. Unfortunately, 2-cycle oil doesn't mix with alcohol the way it does with gasoline. To prevent possible lubrication problems, a vegetable oil such as castor oil should be used.

Straight alcohol can be made to work in a diesel engine, but the high octane and low octane rating of ethanol gives it poor combustion qualities as a compression ignition fuel. Unmodified injector pumps and injectors can be damaged by water in the fuel or by a lack of lubrication. A vegetable based oil must be mixed with the alcohol to provide adequate lubrication unless the injection system is designed to supply its own lubrication.

Straight ethanol does not have to be water free. It can contain up to 20 percent water (160 proof). Fuels containing from 5 percent to 10 percent water (180 proof to 190 proof) actually perform better than anhydrous (200 proof) fuel.

Ethanol is an organic solvent and can cause some plastics to soften. Plastics are sometimes used in carburetor parts or fuel filters. If problems are encountered, metal materials or alcohol resistant materials can be substituted.

Straight alcohol burns cleaner than gasoline and leaves a sweet smelling exhaust. Carbon monoxide and hydrocarbon emissions are decreased substantially and oxides of nitrogen might or might not be decreased. Evaporative emissions might be somewhat higher.

HOW ALCOHOL AND GASOLINE DIFFER

Modifications are necessary to burn straight alcohol because of differences in the physical properties of ethanol and gasoline (Table 12-1). The chemical formula for ethanol is C_2H_5OH. Without getting into a detailed discussion of the chemistry involved, just remember than the one atom of oxygen in the formula makes a big difference in the way the fuel burns. Gasoline, which is a blend of several hydrocarbons (octane, heptane, pentane, etc.), contains no oxygen; only hydrogen and carbon. The difference is that ethanol requires less additional oxygen to burn than gasoline. Engineers have a term for this. They call the amount of air necessary to completely burn a

Table 12-1. Fuel Comparison.

	Ethanol alcohol	Gasoline
Chemical formula	C_2H_5OH	C_xH_x (a blend of several hydrocarbons)
Octane rating (RON)	106	92 (average)
Ideal air/fuel ratio (stoichiometric)	9:1	14.7:1
Energy content of stoichiometric fuel mixture	94.8 Btu/ft^3	94.8 Btu/ft^3
Heat value (Btu/lb.)	12,165 average	19.625 average
Heat value (Btu/gal.)	80,167 average	119,712 average
Latent heat of vaporization	361 Btu/lb.	140 Btu/lb.
Weight per gallon	6.59 lbs.	6.1 lbs. (average)
Boiling point	173° F	100° - 400°F (depends on blend)
Flash point	70°F	−50°F

given quantity of fuel the *stoichiometric ratio*. In other words, the stoichiometric ratio is the ideal air/fuel mixture for complete combustion.

For gasoline, the stoichiometric ratio is about 14.7:1, depending on the blend of hydrocarbons in the fuel. It would take approximately 15 pounds of air to completely burn 1 pound of gasoline under ideal circumstances. The stoichiometric ratio for ethanol is 9:1, or 9 pounds of air for every pound of alcohol.

Beneath the hood, most carburetors are calibrated somewhat richer than this to compensate for a certain amount of incomplete combustion that can't be avoided. But regardless of the exact air/fuel ratio, burning ethanol requires a richer mixture—or roughly 60 percent more fuel.

Maintaining the correct air/fuel ratio for efficient combustion, therefore, involves making some adjustments when switching from gasoline to ethanol. Either the amount of air entering the carburetor must be reached or the amount of fuel flowing through the jets must be increased. Since choking the carburetor restricts power, the main jets must be enlarged 30 percent or 35 percent in diameter. Failure to make this change would create an overly lean condition, resulting in misfiring and a drastic drop in power. It's also doubtful the engine would start with such a lean mixture.

Naturally, increasing the fuel flow through the main jets increases fuel consumption. Based on the difference in air/fuel ratio (14.7 :1 versus 9:1), fuel consumption goes up about 60 percent. But as you will learn in the paragraphs ahead, alcohol has a number of redeeming qualities that can be used to narrow this gap to 50 percent or less.

Btu's Versus Power

Part of the reason an engine takes more fuel when using alcohol is because of a difference in energy content between the two fuels. Gasoline contains more Btu's per gallon than ethanol. In other words, a gallon of gasoline will produce more heat when it is burned than a gallon of alcohol (19,625 versus 12,165 Btu's). Since heat equals energy, the natural assumption is that gasoline produces more power than alcohol, or switching to alcohol would mean a loss in power. Not true. Gallon for gallon, gasoline does produce more power than alcohol. But at the correct air/fuel ratio for each fuel, the total Btu output is nearly identical. This is because there is 60 percent more alcohol in the air/fuel mixture—which exactly offsets the lower heat content of the fuel. The engine uses more fuel, but the net horsepower remains the same.

Heat Of Vaporization

Alcohol has another characteristic that might or might not be considered an advantage over gasoline, depending on the application. Alcohol has a greater *latent heat of vaporization.* This means that it takes more heat to vaporize alcohol than it does gasoline. The amount of heat necessary to vaporize ethanol is 361 Btu's per pound compared to 140 Btu's per pound for gasoline.

When fuel is drawn through the carburetor into the engine, it absorbs heat from the surrounding air as it vaporizes. This effectively cools the air/fuel mixture and when air is cooled it contracts and becomes more dense. This allows more air/fuel mixture to be drawn into the cylinders, which in turn boosts horsepower. Racers have long made use of this characteristic and it is part of the reason why Indianapolis 500 race cars burn alcohol rather than gasoline. Actually they burn methyl alcohol rather than ethanol because methyl has an even higher latent heap of vaporization—474 Btu—pounds—which allows an even denser fuel mixture to be crammed into the engine's cylinders.

The drawback of having a high heat of vaporization is that it requires more heat in the intake manifold to keep the fuel properly vaporized. If there isn't enough heat, the fuel won't vaporize completely. The result is incomplete combustion and wasted fuel.

Cold starting can also be a problem because of the amount of heat needed to vaporize the fuel. Of course, once the engine is allowed to warm up, sufficient heat is generated to keep the fuel vaporized. The cold starting problem begins to make itself known as the outside temperature drops below 50 degrees F.

Fortunately, there are a number of simple modifications that can be used to cure his problem. See the next chapter for details. One method is to preheat the fuel. A small electrical heating element can be installed under the carburetor bowl to heat the fuel so that it will vaporize more easily. Another approach is to squirt gasoline or ether into the carburetor to get the engine going. Yet another is to start the engine on a small auxiliary supply of gasoline and then switch over to alcohol once the engine has warmed up. For prolonged cold weather operation, intake air should be preheated off the exhaust manifold. A fuel line preheater and increased intake manifold heat are also recommended.

Octane Rating

This is where alcohol really comes out on top. Ethanol has a higher octane rating than gasoline; 106 versus 92 for unleaded regular (research octane or RON). The higher octane rating has more compression and ignition advance can be used without fear of spark knock or detonation. This in turn greatly improves the thermal efficiency of the engine.

Detonation is the destructive reaction that occurs when a fuel explodes prematurely or erratically inside an engine. Normal combustion is a smooth, controlled process and is necessary for optimum efficiency and performance. Detonation can be caused by too much compression, over advanced timing or too low an octane in the fuel itself. Anyone who has bought unleaded gas recently has probably experienced the pinging caused by low octane fuel. A little pinging won't cause any noticeable damage. But if it becomes too great, the sharp, hammer-like rapping can crack pistons, break rings or cause premature bearing failure.

Alcohol doesn't suffer from this problem because of the higher octane rating. The typical compression ratio in modern engines is around 8 to 1. With alcohol, the compression ratio can be increased as high as 12 to 1 without fear of detonation. The reason for raising the compression in the first place is to improve the thermal efficiency of the engine. The higher the ratio, the greater the power. It's like compressing a spring; the harder you squeeze it, the harder it snaps back. The same principle applies to the air/fuel mixture inside the cylinders. By squeezing it into a smaller volume, more of the combustion heat can be captured as useful mechanical energy. That is why diesel engines with their super high compression ratios are so fuel efficient. They capture heat energy that would otherwise go out the tailpipe.

Steam Pressure

As mentioned earlier, straight alcohol fuel doesn't have to be water free. The presence of up to 10 percent water in the fuel has a beneficial effect on overall power and fuel efficiency. Water increases power when it is converted to steam.

Combustion temperatures transform the water droplets into vapor. This adds to the pressure already created within the cylinders and gives a little added push against the pistons. But more importantly, the conversion of water into steam consumes heat—at a rate of about 1,100 calories per gram—at a critical instant. The absorption of heat slows the combustion process. Instead of hitting a sharp peak in temperature and then rapidly falling off, the combustion is more gradual. This effectively extends the duration of the burn, which allows more overall pressure to be created for more total power.

The net result, according to a report issued by the U.S. Department of Energy, is that ethanol provides about 5 percent more power than gasoline at equal compression ratios, and 15 percent or more power if the compression ratio is raised to take advantage of the higher octane rating or ethanol.

In terms of fuel efficiency, this means the difference in mileage between gasoline and ethanol is not as great as the 60 percent figure predicted by theory. By raising the compression rate of the engine, advancing the ignition timing and making the appropriate carburetor adjustments, the difference in fuel mileage might actually be 40 percent or less thanks to the improved thermal efficiency of the engine.

Remember too, that if you're making your own fuel the slightly higher fuel consumption is more than offset by the difference in price.

STRAIGHT ALCOHOL FUEL IN A DIESEL

The same qualities that make straight alcohol a good alternative to gasoline make it a poor substitute for diesel fuel. In a diesel engine, there are no spark plugs. The fuel is ignited by the heat of compression along. That is why the compression ratios of diesels are 16 to 1 or higher.

A diesel works as follows: Air is drawn in through the intake manifold (there is no carburetor) and into the cylinders. As the piston approaches top dead center, the force of compression causes the air to become extremely hot—hot enough to ignite any fuel it comes in contact with. At precisely the right moment, the injector sprays a fine mist of fuel directly into cylinder. It strikes the hot air and instantly ignites.

To get this sequence of events working correctly, the injector timing, spray pattern, fuel pressure, fuel mixture and compression ratio must be balanced according to the viscosity of the fuel, its burning properties and cetane rating (ability to ignite under compression). And that's where the problem lies in converting from diesel to alcohol for the do-it-yourselfer.

For starters, diesel oil is a heavy, thick fluid whereas alcohol is light and thin. This leads to immediate problems in the injectors. The heavy diesel oil acts as a lubricant. Without such lubrication, the high working pressures and extremely close tolerances will cause rapid self-destruction. To prevent such calamities, straight alcohol must have a lubricant added to it if it is to be used successfully in a modified diesel engine. Since mineral oil won't mix properly with alcohol, a vegetable oil such as castor oil must be used 5 percent to 20 percent of the fuel total.

Soybean oil is another choice. A diesel engine can be run on straight soybean oil or sunflower oil. The only other option is to design the injection system with alcohol in mind. This involves recalibrating internal tolerances and working pressures to compensate for the thinner viscosity of alcohol and adding an independent lubrication system for the injectors. In short, it takes a whole new injection system and this puts alcohol-fired diesels out of the reach of the do-it-yourselfer in most cases.

There is also the problem of moisture in the fuel. Water and diesel injectors are not meant for one another. The slightest amount of water can cause rust or corrosion that will ruin a set of injectors. This is because clearances are typically on the order of millionths of an inch inside an injector. So you can see it doesn't take must rust to foul things up.

Most diesels have fuel line filters to trap water before it can do any harm. The water is easy to separate because diesel oil and water don't mix. But alcohol and water do mix. They mix together so well it's impossible to separate the two except by distillation. Any water in the fuel, therefore, would be carried right through to the injectors where it would cause problems.

One way to avoid this is to use anhydrous alcohol (200 proof). Removing that last 5 percent water from the alcohol is very difficult for the do-it-yourselfer—about 190 proof is tops for home or farm fuel production. Beyond that, chemical extraction techniques or drying must be used to remove the remaining water. This can be time consuming and expensive. The only other alternative to this route is to install an injection system that will tolerate a small amount of water.

As for the compression ignition qualities of alcohol, the high octane and low cetane rating means that it resists spontaneous combustion. To overcome this resistance, the compression ratio should be increased—probably as high as 24 to 1. However, this is approaching the mechanical limits of the average diesel. At such a high compression ratio, there is very little room left in the combustion chamber. Using a lower compression ratio would compromise the combustion properties of the fuel.

Injector timing must also be advanced to compensate for the slower burning speed of alcohol. Without this modification, the delay in ignition would seriously limit engine power. The best use of straight alcohol is probably in a modified spark ignition engine rather than a diesel.

This isn't to say alcohol can't be used in a diesel, but you would be using the fuel in an application for which it is not well suited. Most do-it-yourselfers would find the performance problems and modifications to be more of a headache than the conversion would be worth.

A retrofit diesel conversion kit could be developed, but the extent and complexity of the modifications would probably make the cost prohibitive. It would probably be easier and more economical to swap engines—that is, replace an existing diesel engine with a gasoline engine modified to burn alcohol.

As for alcohol-powered diesels from the factory, the chances seem remote. The agricultural equipment manufacturers recognize the greater potential of alcohol for replacing gasoline so they've shown little interest in developing diesel engines that would run on alcohol. The only exception to this has been in the country of Brazil where the government has mandated a change to alcohol fuels. Alcohol-powered diesels have been developed there but with limited success. Again, the problem goes back to the basic combustion qualities of alcohol. A better substitute for diesel fuel in this case might be something like soybean or sunflower oil. Either can be burned with very few modifications or problems.

Alcohol does have one important application as far as diesels are concerned. It can be used as a supplemental fuel to boost power or to decrease diesel fuel consumption.

ALCOHOL AS A SUPPLEMENTAL FUEL

To use alcohol as a supplemental fuel in a diesel, a carburetor can be installed on the air intake, or a fuel nozzle can be installed at the turbocharger air inlet. The carburetor or injector nozzle is then adjusted to supply enough alcohol to meet from 10 percent to 50 percent of the fuel requirements of the engine (Table 12-2). Depending on the amount of alcohol entering the engine via the air intake, the diesel injector pump is turned back an equivalent amount to compensate for the extra fuel. This reduces the overall diesel fuel consumption. Engineers some-

Table 12-2. Alcohol Fuel Applications.

SPARK IGNITION ENGINES	10% ethanol/90% gasoline blend (gasohol)	No modifications necessary. Improved performance.
	10% to 20% ethanol blends	Minor adjustments might be needed. Improved performance but reduced mileage.
	100% ethanol	Carburetor and timing modifications required, fuel preheating also necessary for cold weather operation. Improved performance but mileage reduced up to 50%
DIESEL ENGINES	10% to 30% ethanol blend with 90% to 70% diesel fuel (diesohol) Injection system changes may be needed —power reduced	100% ethanol Major engine modifications required. Power reduced (not considered to be a practical use).

times refer to this type of supplemental fuel arrangement as "fumigation."

Alcohol can also be used in this manner to boost horsepower. Under light loads or partial throttle operation, the engine runs on diesel fuel only. But under demand, additional alcohol is injected into the air intake to enrich the mixture and boost power.

True dual fuel diesels are mostly engines that are designed for stationary use, such as for generating electricity and pumping water. Most burn either a mixture of diesel and natural gas or a mix of diesel and propane. Some burn diesel and gasoline. In these engines, the supplemental fuel is fumigated into the air intake to create a very lean mixture. The mixture is so lean that it won't ignite when compressed. This is where the diesel fuel comes into play. When the air/fuel mixture is compressed, a small amount of "pilot" diesel fuel is injected into the combustion chamber. This small charge of diesel fuel ignites instantly and sets off the rest of the mixture. Therefore, most of the power comes from the supplemental fuel and not the diesel fuel.

Alcohol can be used as a supplemental fuel in a gasoline powered spark ignition engine. A nozzle sprays a fine mist of alcohol into the air stream as it enters the carburetor. The nozzle is usually mounted inside the top of the air cleaner directly over the carburetor throat. The carburetor is adjusted to lean the fuel mixture in proportion to the amount of alcohol used. With this technique, alcohol might supply from 10 percent to 25 percent of the total fuel requirements of the engine. This would cut gasoline consumption by roughly the same amount. Using this technique rather than just blending the alcohol directly with the gasoline allows the use of lower proof fuel. Instead of the 195 proof to 200 proof alcohol that is necessary to make gasohol, fuel containing as much as 20 percent water can be used.

Another application of this vapor injection technique is to use only enough "supplemental fuel" to cool and humidify the incoming air. Straight water can be used for this purpose or a blend of half water and half alcohol in winter months (to prevent freezing can be used. Most systems designed for this purpose supply water only when the engine vacuum drops or is under load. The water vapor produces the gradual combustion phenomenon described earlier under straight alcohol fuels. This improves combustion efficiency, reduces detonation and results in more power and better fuel economy. Such vapor injection kits can be made from scratch or purchased for a variety of manufacturers. The trouble is, the main fuel is still gasoline. The supplemental fuel approach only prolongs the dependence of fossil fuels. If at all possible, some means of converting to straight alcohol should be found.

ALCOHOL/GASOLINE BLENDS

Alcohol can be mixed with gasoline in any proportion. The only requirement is that the alcohol be relatively free of water. If the amount of water is more than a small percentage of the blend, the alcohol and gasoline will *phase separate*. The fuels will partially separate into two layers. The top layer will contain gasoline and alcohol and the bottom layer will contain alcohol and water. This can cause hard starting or stalling, especially on cold mornings. During summer months, alcohol and gasoline can be blended successfully if the alcohol is at least 195 proof. In winter months, it should be 198 proof to 200 proof to prevent trouble.

Blending alcohol and gasoline extends the basic fuel supply and improves the octane rating for better performance. Tests have shown that up to 20 percent ethanol can be blended with gasoline before carburetor adjustments become necessary. Beyond that, the jets must be opened up to compensate for the increasing proportion of alcohol. Otherwise the air/fuel mixture will be too lean. This can cause misfirings and a lack of performance. As the percentage of alcohol increases, cold starting also becomes more of a consideration. Additional modifications such as a fuel preheater might be necessary to avoid cold starting problems.

Alcohol blends have been around for a long time. Between World War I and World War II, European countries were faced with a serious shortage of gasoline. As a result the European governments turned to alcohol/gasoline blends to extend their limited fuel supplies.

In Germany, two blends were available. *Monopollin* was a mixture of 25 percent alcohol and 75 percent gasoline, and *Aral* contained 20 percent alcohol, 20 percent benzol and 60 percent gasoline. France adopted a 50/50 blend of alcohol and gasoline, called *Carburant National*, right after World War I. All government departments and Paris buses were required to burn the fuel. Unfortunately, the 50/50 blend in unmodified engines proved troublesome when it came to cold morning starts. In 1931, the French changed the law to include all commercial vehicles and changed the composition of the blend to 25 percent to 35 percent alcohol. The name of the fuel was also changed to *Heavy Carburant National.*

In jolly old England, the problem was not as severe. Nevertheless, two alcohol blends were offered: *Koolmotor*, a blend of 16 percent alcohol and 84 percent gasoline; and *Cleveland-Discol*, a racing fuel composed of 79 percent ethanol, 9 percent acetone and 10 percent gasoline.

Czechoslovakia made the use of alcohol fuel blends mandatory in 1928. The fuel as called *Dynalkol* and consisted of 50 percent alcohol, 30 percent gasoline and 20 percent benzol. The alcohol was itself a blend of 95 percent ethanol and 5 percent methyl alcohol (made from wood).

In other parts of the world, the fuel shortage between the wars was met in similar ways. In South Africa, a mixture of 60 percent ethanol and 40 percent ethyl ether (both made from sugar products) was sold under the name *Natolite*.

In South America, the nations not blessed with petroleum reserves were forced to turn to alcohol also. In Brazil, two blends were offered. One was called *Gasalco* and contained 82 percent alcohol and 18 percent gasoline. The other blend consisted of 70 percent alcohol and 30 percent ether.

In the United States, the "power alcohol movement" achieved brief success in the 1930s. But for the most part, American alcohol blends, *Alcoline* and *Agrol* achieved only limited popularity because inexpensive gas was readily available. Although the agricultural markets were severely depressed at the time and sorely needed new outlets, the alcohol blends just couldn't compete economically. With little incentive to buy the blends and nothing but bad publicity from Big Oil interests, the blends quietly disappeared. Not until the Arab Oil Embargo of 1973, did alcohol/gasoline blends again receive consideration in this country.

GASOHOL

The new alcohol blend to emerge from the latest fuel crisis is "gasohol". In 1975, the state of Nebraska ran a test to find out if motorists would buy the 10 percent alcohol, 90 percent gasoline blend. The state provided a service station in the town of Holdrege with 20,000 gallons of the fuel, which was thought to be enough for a year's supply. It was gone in two weeks.

By law, the alcohol that goes into gasohol must be made from a renewable resource. This means either ethanol fermented from grain, crop or other farm products, or methanol distilled from wood pulp. The law also states that gasohol should contain no more than 10 percent ethanol.

For a number of reasons, ethanol has so far been the only alcohol used to make gasohol. Part of the reason has to do with one of Mother Nature's laws. Excess water can cause gasohol to separate. But because ethanol is less sensitive to this problem than methanol, ethanol makes the better blend. Another reason is that most commercial distilleries are designed to produce ethanol from grains, not methanol from wood pulp. This means that methanol made from natural gas, as most of it is made today, does not qualify for gasohol under the definition "alcohol made from a renewable resource."

Directions For Mixing Gasohol

The following guidelines should be observed when gasohol is mixed:

- Use 9 parts gasoline to every 1 part ethanol.
- Anhydrous ethanol should be used.
- The fuels should *not* be mixed in the fuel tank of the vehicle. Use a separate mixing tank.
- Allow the fuels to mix and to settle. Any water/alcohol that is going to separate will settle to the bottom of the tank.
- The gasohol fuel should be siphoned from the *top* of the tank—not the bottom. This prevents the introduction of excess moisture into the fuel tank of the vehicle.
- The water/alcohol that separates can be drained from the bottom of the mixing tank for redistillation.
- If a lower proof alcohol is used to make gasohol, the mixing ratio should be increased slightly to offset the additional water that will settle in the mixing tank. For example, if 100 proof alcohol (50 percent water) is used, the mixing ratio should be increased to 8 parts gasoline to every 2 parts alcohol. This is because approximately half of the 100 proof fuel will settle out (which will be mostly water).

Gasohol Performance

The biggest advantage gasohol has is that it can be used in any gasoline engine without modification. The higher octane rating of gasohol is equivalent to unleaded premium or roughly 92 pump octane and it offers greater resistance to detonation. As for emissions, preliminary Environmental Protection Agency tests indicate that carbon monoxide emissions will decrease in most cases, but hydrocarbons and oxides of nitrogen might either increase or decrease, depending on how the engine is calibrated. Evaporative emissions might also increase somewhat.

Most people aren't as concerned about emissions as they are about fuel mileage and performance. Gasohol might or might not give them what they're looking for. Some gasohol users claimed big improvements in mileage while others have noted no difference or a slight decrease. One study, the *Nebraska Two Million Mile Gasohol Road Test*, showed an average overall increase in mileage of 5 percent with gasohol.

Mileage and performance differences depend entirely on the application. Because of the chemical composition if gasohol, the 10 percent ethanol typically causes a "leaning effect" in the air/fuel ratio. On cars built in the 1960s with relatively rich air/fuel ratios, gasohol usually improves mileage. But on cars made in 1973 and 1974 with excessively lean calibration, loss of mileage and driveability problems might result (misfiring, hesitation, etc). Cars built since 1975, on the other hand, might or might not pickup some extra mileage and power, again depending on how the engine is calibrated.

On the newer electronic feedback systems, such as GM's C-4 closed loop system or Ford's EEC-II and EEC-III (Electronic Engine Control), an oxygen sensor in the exhaust manifold compensates for any leaning effect by maintaining the correct air/fuel ratio at the carburetor (Fig. 12-1).

Gasohol has also been blamed for some other problems. For one thing, alcohol is a solvent—which means it might soften some plastics found in fuel lines, fuel filters, fuel pumps or carburetors. But at the 10 percent concentration, this has not been a problem. In fact, all the Detroit auto manufacturers have extended their warranty coverage to cover the use of gasohol.

Alcohol might also dissolve some of the varnish that accumulates in the fuel tank, which might in turn clog the fuel filter. Fortunately, such problems have been rare. If it does happen the first time gasohol is used, the cure is simple enough: just replace the fuel filter with a new one. Once usually does the trick.

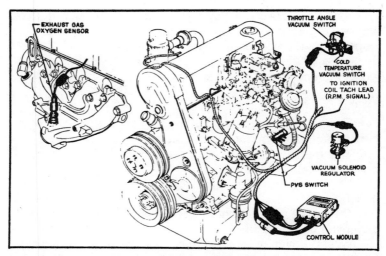

Fig. 12-1. An oxygen sensor in the exhaust manifold of this electronic engine control system compensates for any "leaning effect" due to changes in the air/fuel ratio.

GASOHOL TEST RESULTS

The following is a summary of the test results recorded by the state of Nebraska during an exhaustive evaluation of gasohol. A fleet of state-owned vehicles was used to accumulate over two million road miles with gasohol fuel. The vehicles were used in normal everyday work and were driven under all types of weather conditions. Because this was a "real world" test and not a laboratory experiment, the results can be considered valid for what the average motorist might receive from gasohol.

Fuel Economy

In 99 percent of the cases, cars fueled with gasohol got more miles per gallon than the cars fueled with unleaded gasoline.

The gasohol fueled test cars got an average of 5 percent more miles per gallon than those on gasoline.

The thermal efficiency (miles per million Btu's) of gasohol is about 9 percent greater than for unleaded gasoline.

Engine Wear and Deposits

Micrometer measurements of engine cylinders indicate no difference in wear between gasohol and unleaded gasoline.

Inspections of spark plugs, valves and valve seats indicate the use of gasohol does not reduce any more wear or carbon deposits than unleaded gasoline.

No premature failure of any engine part or fuel line component was encountered during the test.

Vehicle Performance

The higher octane rating of gasohol eliminated pinging in cars that suffered from detonation on unleaded gasoline.

The gasohol fuel test cars appeared to drivers to have better pick-up and power.

No vapor lock was ever reported for test cars using gasohol.

Drivers of gasohol cars experienced no hard starting or stumbling of the engine.

Overall driver satisfaction with the performance of vehicles fuel with gasohol was high.

Exhaust Gas Composition

The exhaust from gasohol fuel contained an average of 30 percent less carbon monoxide than that from unleaded gasoline.

The NO_x (oxides of nitrogen) and unburned hydrocarbons from both fuels was the same.

The total emissions from gasohol fuel were slightly less than from unleaded gasoline.

Gasohol has no effect on the catalytic converter.

Blending and Storage of Gasohol

No difficulties were reported in blending anhydrous grain alcohol (ethanol) with unleaded gas to make gasohol.

No difficulties were experienced in transporting gasohol from the refinery to the filling stations.

No phase separation due to water accumulation was never identified in any of the storage tanks or test vehicles.

Other Test Results

An American Automobile Association test showed that 88 percent of the vehicles tested ran as well or better on gasohol. The 12 percent that didn't run as well could have been adjusted to compensate for the "cleaning effect" created by the 10 percent ethanol in gasohol.

In Illinois, a test of approximately 1800 state owned vehicles yielded positive results in each category tested for gasohol fuel. As a result, all state vehicles now use the fuel.

The Iowa Development Commission (a state agency) conducted a 90-day gasohol marketing trust from June 15, 1978 to September 15, 1978, in which 232,608 gallons of gasohol were sold through five service stations. The results showed 67 percent of the users reported improved performance, 29 percent cited improved mileage, three out of every five were repeat customers, 90 percent of the users said they would purchase gasohol if it were available at most stations, and last but not least, gasohol outsold unleaded regular nearly four to one.

The Standard Mutual Insurance measured mileage and vehicle acceleration of vehicles fueled with gasohol. Their test results showed an average increase in mileage of one mile per gallon and an improvement of three-fourths of a second in 0 to 50 mph acceleration.

Even Ma Bell has gotten into the act. A telephone company test fleet showed an average increase of 3.03 percent in mileage, a noticeable savings in engine maintenance, and the elimination of gas line freezing problems during the winter months. Note the gas line anti-freeze you dump in your gas tank is nothing more than methyl alcohol and ethanol is even better than methanol when it comes to absorbing water.

Other Alcohol/Gasoline Fuel Blends

Another alcohol/gasoline blend to hit the market is a blend called *HydroFuel*. The main difference between hydrofuel and gasohol is that the alcohol is not anhydrous. It contains up to 5 percent water. Alcohol and gasoline do not separate in hydrofuel because a "coupling gent" is added to keep the two together. The chemical is called *Hydrelate*, and is available from the United International Research Co. Inc., of Hauppauge, New York. It sells for approximately $6.00 per gallon and can be used by anyone who wishes to blend lower proof alcohol with gasoline. The recommended proportions are 10 percent 190 proof ethanol, 89 percent unleaded gasoline and 1 percent hydrelate.

DIESOHOL

Alcohol can also be blended with diesel fuel in almost any proportion, but unfortunately the results are not the same as with gasoline blends (Figs. 12-2, 12-3 and 12-4).

In an unmodified diesel engine, increasing the amount of alcohol in the fuel reduces engine power, increases overall fuel consumption per unit of work, delays combustion and increases engine noise. These are the conclusions reached by a team of researchers at the University of Minnesota. Alcohol isn't a good compression ignition fuel. The more the diesel fuel is diluted with alcohol, the worse will be the results. Adjustments can be made to compensate for the differences somewhat, but diesohol blends have not been very successful.

According to the test results, blends from 10 percent ethanol up to 40 percent ethanol were tried in several diesel

Fig. 12-2. A single cylinder diesel at General Motors Research Laboratories used to determine the combustion properties of various diesohol blends. Tests are being conducted in three ways: (A) Ethand is being blended with No. 2 diesel oil; (B) The alcohol is put into the engine immediately before the turbocharger. This eliminates the need for anhydrous alcohol and lowers engine temperatures; and (C) the alcohol and diesel fuels are kept in separate tanks and mixed just prior to entering the injector. One promising technique is to use a separate injector for each fuel—one for the alcohol and one for diesel.

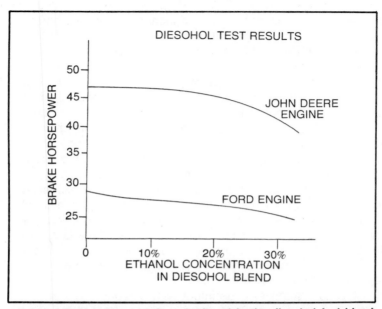

Fig. 12-3. As the concentration of ethanol in the diesohol fuel blend increased, the horsepower fell off for both test diesel engines.

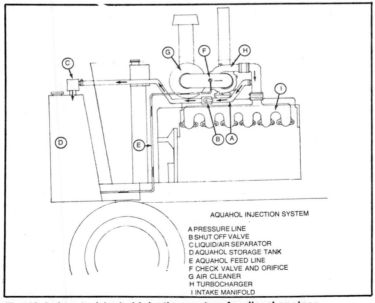

AQUAHOL INJECTION SYSTEM

A PRESSURE LINE
B SHUT OFF VALVE
C LIQUID/AIR SEPARATOR
D AQUAHOL STORAGE TANK
E AQUAHOL FEED LINE
F CHECK VALVE AND ORIFICE
G AIR CLEANER
H TURBOCHARGER
I INTAKE MANIFOLD

Fig. 12-4. A water/alcohol injection system for diesel engines.

engines. But as the percentage of alcohol increased, horse-power decreased. This was attributed to the lower Btu content of alcohol and delayed ignition.

Anyone attempting to mix diesohol will also discover another problem—phase separation. Diesel fuel is even less tolerant of moisture than gasoline. This means the slightest trace of water will cause the two fuels to separate. The amount of water that can cause this is less than 1 percent. The problem also seems to be worse with 10 percent diesohol blends rather than 20 percent mixtures. Perhaps this is because the larger proportion of alcohol will "hide" the moisture better.

Another possibility to consider is injector damage. Just as with using straight alcohol in a diesel, using a blend of alcohol can reduce the lubrication or contaminate the internal injector parts with moisture. Most agricultural equipment manufacturers don't recommend the use of diesohol blends for this very reason. And if damage results, you might have a hard time collecting on your warranty coverage.

If you're going to experiment with diesohol, use an old tractor that can be repaired cheaply in case something goes wrong. Without modifications to the injection system, diesohol blends over 10 percent are not recommended.

Chapter 13
Engine Conversion

When converting an engine from gasoline to alcohol, two basic modifications are required to compensate for the different fuel characteristics of alcohol:

—The main fuel jet(s) in the carburetor must be enlarged 30 percent to 35 percent in diameter.

—The ignition timing should be advanced 4 degrees to 6 degrees over the stock setting.

The carburetor jets must be enlarged to produce the richer air/fuel mixture alcohol requires. Increasing the diameter of the jet orifice 30 percent to 35 percent increases the fuel flow through the jet about 60 percent. That is just the right amount to produce the correct air/fuel ratio. Failure to make this change will starve the engine for fuel. The resulting overly lean condition will cause misfiring, hard starting and a drop in overall power.

The ignition timing is advanced to compensate for the slower combustion speed of alcohol and higher octane. Advancing the timing also improves fuel economy and power.

There are additional modifications that can be made to improve performance, fuel economy and cold weather startability. Whether or not you choose to make any of these modifications will depend on the climate where you live, what (if any) problems you encounter with your conversion and how intent you are upon getting the most out of your homemade fuel.

CHANGING THE FUEL MIXTURE

To enlarge the main fuel jets(s), you will have to partially disassemble the carburetor (Figs. 13-1, 13-2 and 13-3). Fear not, for in most cases only the upper half of the carburetor body needs to be removed to reach the jets. Because of this, most of the plumbing, linkage and lower half of the carburetor can be left in place. If the carburetor is removed from the engine for any reason, the gasket underneath it should be replaced with a new one to avoid vacuum leaks. Be sure to use gasket sealer when reassembling or reinstalling the carburetor.

With the top of the carburetor off, you should find the jets somewhere in the fuel bowl. There will be one metering jet for each throttle bore or "barrel." If you are not sure where or what to look for, refer to an engine manual and look up an exploded diagram for your particular make and model carburetor.

Once you have pinpointed your fuel jets, remove them using a screwdriver. Note: some carburetors use fuel metering rods instead of jets. To change the air/fuel ratio, smaller rods can be installed or the outside diameter of the rod tips ground down to increase the fuel flow.

With jets in hand, you can now do one of two things. You can drop by your friendly auto parts or tractor supply store to see if they carry replacement jets that will fit your carburetor. If they do, buy ones that are 30 percent to 35 percent large in diameter. Your other choice is to drill out the old jets yourself.

To determine the "before" size of the jet orifice, gently insert various drill bits into the hole until you find one that just fits. (Don't force the bit or you will get a false reading. Measure the bit with a micrometer or drill gauge and multiply the result by 1.3 and 1.35. Choose the drill bit size that comes closest to either figure. This should be the correct size to drill the hole. If you make a mistake and enlarge the hole too much, just solder the jet shut and redrill with a smaller bit.

To determine whether or not you have enlarged the jets the right amount, check the tailpipe once the engine is fully warmed on alcohol fuel—not gasoline. If black soot coats the inside of the exhaust pipe or is blowing out the pipe, the mixture is too rich. Check the choke to see that it is working properly. It should be wide open once the engine is warm. Check to see that nothing is obstructing the air cleaner. If everything is okay, you'll have to redrill the jets the next size smaller. A chalk white deposit in the tailipe indicates exactly the opposite—too lean an

Fig. 13-1. To burn straight alcohol fuel, the carburetor must be modified to produce a richer air/fuel mixture. This involves partially disassembling the carburetor to change the main fuel metering jet (courtesy Mother Earth News).

air/fuel mixture. In this case, you need more fuel so open the jets up the next size larger. Remember that a little drilling goes a long way so don't overdo it.

Another way to check the mixture is to shut the engine off and remove one of the spark plugs. If the plug tip has black sooty carbon deposits, the mixture is too rich. Decrease the jet size. If the tip has white glazed deposits or a yellowish appearance, the mixture is too lean and more fuel is needed. If all systems are working properly and the carburetor is supplying the correct balance of air and fuel, the plugs will show a light tan to grey coloration.

FINE TUNING THE CARBURETOR

Another adjustment that will have to be made is to enrich the idle fuel mixture. Fortunately, this is relatively simple. All you have to do is find the idle fuel mixture screws(s), and back the screw out one-half or more turns until the engine idles smoothly. This adjustment is necessary because the main fuel metering circuit in the carburetor only controls the air/fuel mixture at part and full throttle. At idle, the idle circuit handles the fuel blending job.

On most of the automobiles since 1973, the idle screw has a limiter cap. The purpose of this cap is to prevent tampering that might adversely affect engine emissions. Since the limiter cap won't allow the screw to be turned out far enough to compensate for the richer mixture, you will have to either pull it off with a pair of pliers or break the limiting tabs off. It's against the law for a garage or service station mechanic to do this for you, but there's no law against doing it yourself. Besides, the alcohol will burn cleaner than the gasoline ever could—even with the pollution controls.

Now that you've taken care of the main and idle fuel circuits, you might have to make some adjustments in the acceleration circuit of the carburetor. This is the system that squirts additional fuel into the carburetor throat when you work the throttle linkage. Its purpose is to prevent hesitation when the throttle is suddenly opened. If you find a hesitation problem,

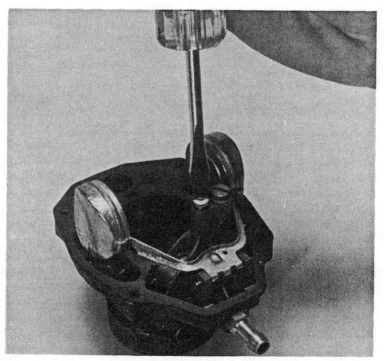

Fig. 13-2. After removing the top of the carburetor, the main fuel metering jet can be removed for modification or replacement (courtesy of Mother Earth News).

Fig. 13-3. The fuel metering jet is drilled 30% to 35% larger to increase the size of the orifice. This will produce the correct air/fuel ratio for straight alcohol fuel (courtesy of Mother Earth News).

simply lengthen the accelerator pump control rod for more travel or drill out the discharge nozzles 30 percent to 35 percent. This will provide a bigger squirt of fuel and hopefully solve the problem.

CHANGING TIMES

Look up the stock ignition timing for your engine in the owner's manual or a shop manual and then increase the initial setting by 4 degrees to 6 degrees. Follow the manufacturer's recommendations for adjusting the timing for correct idle speed, vacuum advance temporarily plugged and points set at the proper gap.

Hook up a timing light and start the engine. Loosen the bolt that holds the distributor in place and slowly turn the distributor housing until the desired timing marks line up. To advance the timing, turn the distributor the opposite way the rotor travels. Don't over advance the timing or you're apt to produce a "pinging" when the engine is accelerated under load. A little fine tuning should achieve setting that gives you the best performance and economy.

COLD STARTING

Unfortunately, alcohol has a reputation for being a hard starting fuel when the outside temperature drops below 50 degrees F. And the colder it gets, the harder it can be to get the engine going. The reason this happens is because alcohol doesn't vaporize as rapidly as gasoline at lower temperatures.

One way to overcome this tendency is to use a "booster fuel" to give the engine a little extra kick on cold mornings. Gasoline or ether can be squirted into the carburetor to create enough fuel vapor to support combustion. Once the engine starts, sufficient heat will be generated to vaporize the incoming alcohol fuel.

A simple booster system (Fig. 13-4) can be made by converting the existing windshield washer system to this purpose or by installing some type of pump and squirter for a small auxiliary tank of gasoline. All that's needed is a means of getting the booster fuel into the intake manifold. The best way to do this is to simply squirt it down the throat of the carburetor. Such a booster system works best if the automatic choke is replaced with a hand-operated manual choke.

A cold start sequence goes like this: With the choke wide open, squirt a small quantity of gasoline or ether down the throat of the carburetor. Then close the choke and crank the engine. If it fails to start the first time, repeat the process. Be careful not to flood the engine.

DUAL FUEL STARTING SYSTEM

Another approach to cold starting, especially in bitter cold weather, is to start the engine on gasoline supplied by an auxiliary fuel system (Fig. 13-5). After the engine has run a few minutes or reached operating temperature, the auxiliary fuel supply is switched off and the alcohol supply is switched on.

To make such a system, a T-value is installed in the fuel line where it enters the carburetor. One line supplies alcohol from the main fuel tank and the other gasoline from a small auxiliary fuel tank. A fuel shut-off valve is installed in each line.

To start the engine, the alcohol line is shut off and the gasoline line opened. An electric fuel pump supplies fuel pressure to feed the gasoline to the carburetor. After the engine has warmed sufficiently to run on the alcohol, the gas line and pump are shut off and the alcohol line is opened.

FUEL PREHEATER

Another technique that can be used to ease starting in cold weather is the use of a fuel bowl preheater (Fig. 13-6). A small electrical heating element made from electrical resistance wire or an electric choke heating element is mounted inside the fuel

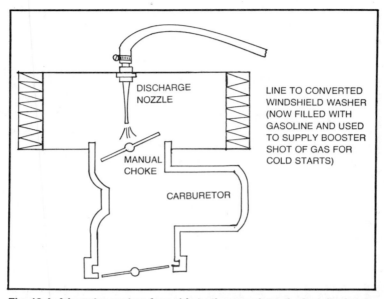

DISCHARGE NOZZLE

LINE TO CONVERTED WINDSHIELD WASHER (NOW FILLED WITH GASOLINE AND USED TO SUPPLY BOOSTER SHOT OF GAS FOR COLD STARTS)

MANUAL CHOKE

CARBURETOR

Fig. 13-4. A booster system for cold starting consists of a fuel discharge nozzle positioned over the carburetor throat. Gasoline is stored in a converted windshield washer reservoir. The old washer line is disconnected from the windshield nozzles and routed to a fitting on the air cleaner. To start the system, the manual choke is opened and several squirts of gasoline are sprayed into the carburetor. The choke is closed and the engine is cranked to start. Once the engine starts, vaporization of the alcohol fuel should not be a problem. The original windshield washer system can be converted to this purpose or a second system can be rigged using an additional fluid reservoir and pump motor. A dash switch would then be used to control the booster squirt of gasoline.

bowl or just underneath it. The idea is to preheat the fuel in the carburetor so that the fuel will vaporize more easily. The fuel bowl preheater can be used for starting or cold weather driving, if the need arises. The only precaution is not to overheat the fuel. If the alcohol gets hotter than 173 degrees F, it will start to boil inside the carburetor. This causes vapor lock, and until the fuel is allowed to cool somewhat, the engine will be next to impossible to start. If the carburetor preheater is used for starting only, usually 15 seconds to 30 seconds of heating will do the trick.

Cold Weather Fuel Line Preheater

Another weapon in dealing with cold weather is the fuel line preheater Figs. 13-7 and 13-8. Again, the idea is to warm the fuel so that it will vaporize more easily. With this method, the fuel

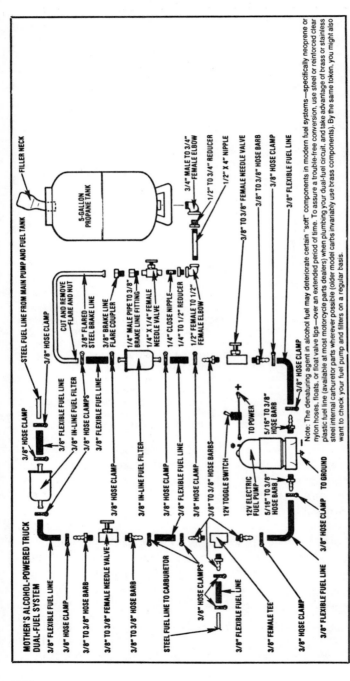

Fig.13-5. Dual fuel starting system as used by the Mother Earth News, alcohol-powered truck. The auxiliary fuel tank (5 gallon propane bottle) is filled with gasoline for cold weather starts. Once the engine is warmed, the engine is switched over to alcohol.

is heated before it enters the carburetor. This system is ideal for prolonged cold weather operation where sub zero temperatures make complete fuel vaporization a real challenge. The fuel line preheater can be made by wrapping electrical resistance wire around a length of tubing. However, the easiest system to install is one that uses heat from the cooling system of the engine. A small section of heater hose is removed and replaced with a length of straight metal tubing. Then a length of metal fuel line is wrapped around it to form a simple heat exchanger. Next, the section is wrapped with aluminum foil and tape for insulation. The fuel line from the heat exchanger to the carburetor is also insulated to prevent heat loss.

Preheating Intake Air

Another fuel system modification that will aid cold weather driveability is the use of a warm air intake system (Fig. 13-9).

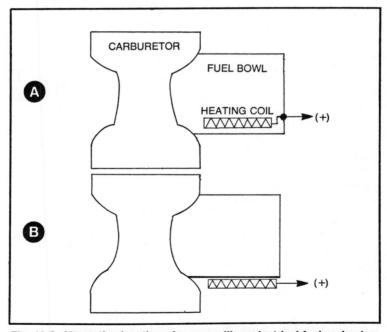

Fig. 13-6. Alternative locations for an auxiliary electrical fuel preheater: (A) epoxy heating coil to bottom of fuel bowl inside carburetor or (B) epoxy the coil to the outside of the carburetor just under the bowl. The heating element can be made from resistance wire or a converted heating element from inside an electric choke. A dash switch is used to turn the fuel preheater on for 15 to 30 seconds prior to starting. Once the engine is running, the fuel preheater is turned off.

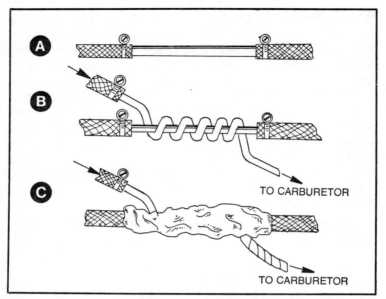

Fig. 13-7. To make a fuel line preheater: (A) replace a section of the heater hose with straight pipe; (B) wrap the fuel line around the pipe section to make a simple heat exchanger; (C) insulate the heat exchanger with aluminum foil and tape, insulate the fuel line to the carburetor to prevent heat loss.

Incoming air is warmed by passing over the hot exhaust manifold before it enters the carburetor. This helps improve fuel vaporization for better performance and economy. Most automobiles since about 1972 use such a system to speed engine warm up, improve cold weather driveability and control emissions. If the engine is not already equipped with such a system, one can be easily fabricated with some ducting and sheet metal. Intake air preheating should be necessary if the outside temperature is less than 50 degrees F. Preheating the air in warm weather can cause an overly lean condition and associated problem.

OTHER FUEL SYSTEM MODIFICATIONS

In some instances, alcohol will react with certain plastic causing the materials to soften or swell. Most automotive plastics are fairly resistant to alcohol. But with the tremendous variety of materials used, a few do-it-yourselfers will be unlucky enough to find some plastics that won't stand up.

Neopreme tipped fuel inlet valves in the carburetor might swell and stick shut. If this happens, replace the valve with one made from solid steel. Troublesome plastic carburetor floats can be replaced with ones made from brass or other floats that are resistant to alcohol. Clear plastic fuel filters might also cause problems. If softening is noted, replace the filter with one that has a metal housing.

INTAKE MANIFOLD HEAT

Alcohol requires plenty of heat to vaporize completely. On most V-8 and V-6 engines, a device known as a *heat riser valve* is installed on one of the exhaust manifolds (Fig. 13-10). This valve remains closed while the engine is cold and forces the exhaust gas to flow through a channel under the intake manifold to the opposite side. This speeds engine warm up and helps the fuel vaporize in cold weather. You might want to change the rate at which it opens to improve cold weather operation. You can do this by bending the bi-metallic spring slightly to increase spring tension.

On in-line engines, intake manifold heat is supplied by water passages or by positioning the intake manifold next to the exhaust manifold. To get more heat exchange between the intake and exhaust, some type of insulation can be loosely wrapped around the two to trap heat inside. Aluminum foil is probably the best since it is easily molded and resists the high exhaust temperatures.

IGNITION MODIFICATIONS

Because alcohol burns cooler than gasoline, the standard spark plugs might develop a tendency to foul. If they do, the cure is to install the next hotter heat range of plug.

Fig. 13-8. Wrap the fuel line to prevent heat loss once the fuel has passed through the heat exchanger.

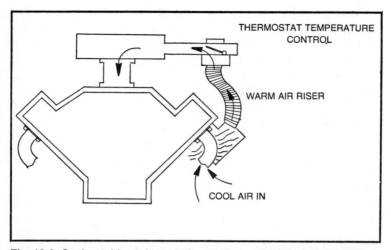

Fig. 13-9. Cool outside air is routed over the hot exhaust manifold to be heated. The warm air is then drawn up through the plumbing and into the carburetor. The warm air speeds engine warm up and provides for better vaporization of the fuel. For cold weather operation on alcohol fuel, the thermostat should be disconnected so that all incoming air will be preheated.

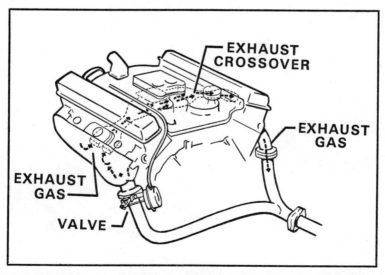

Fig. 13-10. To increase intake manifold heating for faster warm up and improved fuel vaporization in cold weather, the heat riser valve can be modified to open more slowly or only partially. This routes more hot exhaust gas under the intake manifold to preheat the incoming air/fuel mixture.

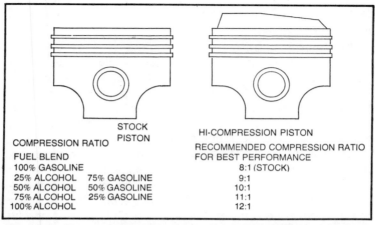

COMPRESSION RATIO		
STOCK PISTON		HI-COMPRESSION PISTON
FUEL BLEND		RECOMMENDED COMPRESSION RATIO FOR BEST PERFORMANCE
100% GASOLINE		8:1 (STOCK)
25% ALCOHOL	75% GASOLINE	9:1
50% ALCOHOL	50% GASOLINE	10:1
75% ALCOHOL	25% GASOLINE	11:1
100% ALCOHOL		12:1

Fig. 13-11. The compression ratio can be increased in relation to the amount of alcohol in the fuel. The increases are made possible because of alcohol's higher octane rating. Compression can be raised by replacing the stock pistons with hi-compression pistons or by milling metal off the head surface to decrease the combustion chamber volume. Piston replacement is recommended over the latter since higher ratios are possible.

COOLING SYSTEM MODIFICATIONS

If driveability problems indicate the alcohol fuel isn't being heated enough to vaporize properly, the cooling system thermostat can be changed to a higher temperature. Use a 190 degree F thermostat to replace the standard one if it's a 160 degree F or 180 degree F unit.

INCREASING THE COMPRESSION RATIO

Although by no means necessary, the compression ratio (Fig. 13-11) is one of the best ways to take advantage of the high octane rating of alcohol. Increasing the compression increases power and fuel economy and generally makes the engine more efficient. With straight alcohol fuel, the compression can be raised to 12.5:1 without danger of detonation.

There are two ways to mechanically increase the compression ratio. One is to install high compression pistons (identified by raised domes on top). The other way is to decrease the combustion area in the head by milling metal off the surface of the head. Milling only goes so far, however, and alignment problems might result between the head and manifolds of V-8 and V-6 blocks. The recommended procedure, therefore, is to install the "pop up" pistons. Replacement pistons can be ordered through a variety of channels, including a local auto parts store or speed shop.

OTHER MODIFICATIONS

Other modifications that aren't absolutely necessary but might be desirable include the installation of a larger fuel tank for increased range. Fuel mileage isn't as good with straight alcohol as it is with gasoline so a large capacity tank might be a good investment. A high percentage of water in the fuel can cause some corrosion problems in a bare steel tank. If rust becomes a problem, a stainless steel, fiberglass or plastic tank that is resistant to alcohol should be installed.

Another modification that involves little work but can prevent big headaches is to clearly label all fuel tanks and fuel filler caps on converted vehicles "Alcohol Fuel Only." Filling up with wrong fuel can lead to total confusion as to why the engine won't run correctly. After an engine has been modified to run on straight alcohol, it will run too rich on gasoline.

Of course, the nice thing about do-it-yourself conversions is you can always undo them if, for some reason, you decide to switch back to gasoline. If you keep the modifications fairly simple, the costs and effort should be minimal. Any do-it-yourselfer with a little mechanical skill and knowledge should be able to complete a simple conversion in an afternoon.

Diesel Conversions

Because of the problems outlined in the previous chapter, diesel conversions are probably beyond the abilities of most do-it-yourselfers. Rather than converting to straight alcohol, using alcohol as a supplemental fuel seems to produce better results. This can be accomplished by adding a carburetor to the diesel air intakes or a fuel nozzle to the air intake just ahead of a turbocharger. The section that follows describes one such system that is offered as a kit.

Water/Alcohol Injection System For Diesels

The *M&W Gear Co. of Gibson City, Illinois* has developed an alcohol injection system that will boost horsepower and reduce diesel fuel consumption nearly 25 percent (Fig. 13-12). The system is also fully automatic.

The injection of a mixture of water and alcohol into the intake airstream of a diesel engine under load increases the horsepower output, reduces intake and exhaust temperatures

Fig. 13-12. A water/alcohol injection system can boost diesel horsepower and reduce diesel fuel consumption. This system is a test rig used by the M&W Gear Company, Gibson City, Illinois.

and prolongs engine life. The reduction in the use of a diesel fuel alone make the use of such a system worth considering.

The water/alcohol injection system consists of a tank containing a 50/50 mixture of water and ethanol, a pressure line and a feed line. As the load on the engine increases, turbocharger boost pressure increases. The alcohol tank is pressurized with turbocharger air at 5 pounds of boost. This forces the mixture through the feed line to the intake side of the tubocharger. As the boost increases, more of the water/alcohol mixture is injected automatically. An orifice in the feed line meters the mixture. The size of this orifice is determined by the requirements of the engine.

As the mixture is injected into the airstream, the air is cooled, providing a denser charge of air for combustion. The air/water/alcohol mixture then combines with diesel fuel in the combustion chamber to create a cooler, more powerful burn.

The cylinder sleeves pistons and valves are said to stay leaner and last longer with the system. Less heat is transferred to the cooling water and oil and cleaner combustion at a time when blow-by is greatest to keep the oil cleaner. Exhaust temperatures are also said to be lower.

According to tests, the M&W injection system reduced fuel consumption from 8½-gallons per hour to 6 gallons per hour while maintaining an even 125 horsepower. Therefore, water/alcohol injection reduced the base fuel consumption 2½-gallons of deisel fuel per hour.

Glossary

alcohol. The name commonly used when referring to ethanol, methanol or any of the other alcohols. Technically speaking, the alcohols are a group of chemically similar organic compounds with the general formula $C_nH_{2n+1}OH$ where C stands for Carbon atoms, H for Hydrogen and O for oxygen. The lower case "n" represents the number of atoms. Alcohols are typically clear liquids, flammable and strong organic solvents. They will dissolve a great many substances, including some plastics.

anhydrous. Without water. When referring to ethanol alcohol, it means the same thing as 200 proof.

ATF. Abbreviation often used for the U.S. Bureau of Alcohol, Tobacco and Firearms. It is part of the Department of the Treasury and is responsible for administering federal laws and regulations governing the taxation, production, denaturation and distribution of ethanol alcohol.

batch. An amount of mash produced for one run through the distillation process.

beer. The liquid portion of the mash containing water, ethanol alcohol and other fermentation by-products. The beer is what is left after straining the mash. The beer is then distilled to separate the alcohol.

biomass. Any plant matter, including grains, vegetables, crop residues, leaves and stalks. When plants grow, they produce biomass from solar energy. The cellulose, starches and sugars produced can then be harvested and fermented into alcohol or methane.

brew. Means the same thing as mash. It is the mixture of grain (or corn etc.), water and yeast that is used to make alcohol.

Btu. An abbreviation for British thermal unit. This a measure of an amount of heat. One Btu equals the amount of heat required to raise the temperature of one pound of water one degree Fahrenheit. When referring to motor fuels, the heat value per gallon or per pound is often expressed in Btu's.

cellulose. The main component of all plant fiber, consisting mostly of long chains of glucose (sugar) molecules. Wood is largely cellulose as is paper, straw, hay, manure and most crop residues. You can burn it for its heat value, roast it in the absence of oxygen to make charcoal or methanol alcohol, or break apart the sugar molecules with enzymes and ferment it into ethanol alcohol. Although people can't digest cellulose, cattle, sheep, and goats can.

condenser. The part of the distilling apparatus that causes the alcohol vapors and steam to condense into liquid. It is located atop the distilling column or still. The hot vapors flow through the condensr and are cooled by outside air or a stream of cold water. A simple condenser can be made from a length of coiled copper tubing.

denature. The process of making ethanol unfit to drink by adding kerosene, gasoline or methanol according to approved formulas by the U.S Bureau of Alcohol, Tobacco and Firearms. All ethanol for non-beverage use must be denatured or else it will be subject to a federal excise tax of $10.50 per 100 proof gallon.

diesohol. The name given any mixture containing diesel fuel and alcohol (ethanol or methanol). Mixtures can range from 10 percent up to 50 percent or more alcohol, but 20 percent ethanol and 80 percent diesel seems to be the most workable blend. Diesohol has a lower viscosity, less lubricity, less heat value and burns slower than straight diesel fuel. Its use is still considered experimental as some injection pumps might fail due to an adverse reaction between the alcohol and certain gasket materials.

distillation. The process used to separate the alcohol from the beer or liquid portion of the mash by vaporization. Since alcohol vaporizes at a lower temperature than water, the beer is heated so that more alcohol is boiled off than water. The vapors then rise into a distilling column for further separation or are funneled into a condenser and made into liquid again. Depending on the equipment used, the alcohol produced might range in proof from less than 100 to as high as 190.

distillation column. The name given to the piece of equipment which separates the water and alcohol vapors. The column usually consists of a tall pipe with internal baffles, screens or packing. As the vapors rise, they condense and re-evaporate with the alcohol rising further than the water. The result is usually high quality (190 proof) alcohol.

distillery grain. Also known as distillers dried grain and solubles (DDG or DDGS), it is the leftover portion of the mash after the water and alcohol have been removed. Think of it as either fermentation residue or a distillation by-product. Depending on what was originally used to prepare the mash, the DDG can usually be used as a high protein food supplement for livestock. For example, each bushel of corn after distilling leaves 17 to 18 pounds of high protein DDG.

DSP. This is what the ATF calls an ethanol alcohol still or production facilities. If you don't have their permission to operate a DSP, they call it illegal.

enzyme. A catalyst produced by a living organism that is used to break down starches into simple sugars, which then can be fermented to make alcohol. Enzymes can be extracted from sprouted barley, corn, etc., or purchased from commercial suppliers. The two classes of enzymes useful for mash preparation are Amylases for starches and Cellulases for cellulose.

ethanol. The alcohol obtained from the fermentation of sugars and starches. Its chemical formula is C_2H_5OH. Also known as ethyl alcohol, grain alcohol, beverage alcohol, distilled spirits or alcohol. It is an excellent motor fuel that is clean burning and has a high octane rating. When used straight, 160 proof to 180 proof seems to work best. When mixed with other fuels such as gasoline or diesel, 200 proof should be used to avoid separation problems caused by water.

ethyl alcohol. Another name for ethanol. The word "ethyl" refers to

Glossary of Automotive Terms

Glossary

Ethyl Alcohol — Gasoline

ethyl alcohol — The first portion of ethanol's formula, which is C₂H₅. Ethyl alcohol has nothing to do with ethyl gasoline, which is a high octane fuel with tetraethyl lead as an additive.

excise tax — The tax imposed by the Federal government on distilled spirits (ethanol for use in alcoholic beverages). The government has the constitutional right to impose taxes on the production, sale or consumption of any given commodity it chooses. Excise taxes have always been controversial. The tax on alcohol has been credited with making moonshining a lucrative business (as well as a crime). When Prohibition was repealed in 1933, the federal excise tax was $1.10 per 100 proof gallon. It is now $10.50 per 100 proof gallon. To avoid this tax, your alcohol must be denatured.

fermentation — The process whereby sugar is converted into alcohol and carbon dioxide by yeast. Nearly any organic material can be fermented, provided starches or cellulose are first broken down into sugars the yeast can feed on. Enzymes can be used to accomplish this and prepare the mash. As the yeast multiply and grow in the mash, they excrete ethanol and carbon dioxide as metabolic wastes. The yeast continue to grow until either most of the available sugar has been consumed or until the concentration of alcohol and CO₂ becomes toxic. When the mash ceases to bubble, fermentation has ended. The time depends on the strain of yeast, the mash, temperature, etc., but generally 80 degrees F to 90 degrees F for 48 to 72 hours will do the trick.

fermentation lock — A simple device that allows CO₂ to escape from the fermentation tank but prevents airborne organisms from contaminating the mash. Water in a double elbow joint acts as a trap, similar to the double joint in the drain plumbing under a sink, to seal off the pipe. Gas pressure within the fermentation tanks escapes by bubbling through.

fumigation — The process of feeding a vaporized fuel into the air intake of a diesel engine or into the air stream above a carburetor. Also called vapor injection. Straight alcohol can be used, straight water or any combination thereof. The technique is used to increase horsepower, reduce the fuel consumption or reduce detonation.

gasohol — Contrary to popular misconception, gasohol can be any blend of gasoline and alcohol (either ethanol or methanol) in any proportion. However, the most commonly agreed upon definition is 10 percent anhydrous (water free) ethanol and 90 percent unleaded gasoline. This is the blend approved by new car warranties and currently available in the United States. The reason for using gasohol is to extend existing gasoline supplies as well as reduce dependence on foreign oil. By law, the alcohol that goes into gasohol must be made for a renewable resource to qualify for the special tax exemptions now being offered by the Federal government and certain states. This means either ethanol fermented from grain, corn or other farm products (or municipal wastes) or methanol distilled from wood pulp. It does not include ethanol made from coal or natural gas. Gasohol typically has an octane rating equivalent to unleaded premium and produces a slight leaning effect on the air/fuel ratio—which might or might not improve gas mileage, depending on a number of factors.

gasoline — A mixture of various liquid hydrocarbons derived mainly from crude oil and used as a motor fuel for internal combustion engines. A non-renewable resource which requires massive investments of capital to find, transport, refine and use. Other than that, it's a great motor fuel.

grain alcohol. Another name for ethanol, ethyl alcohol, beverage alcohol or distilled spirits.

hydrocarbon. Any molecule consisting of a long chain of carbon and hydrogen atoms. Gasoline is a blend of various hydrocarbons such as octane, hexane, heptane, decane and pentane.

hydrometer. An instrument used to measure the specific gravity of a liquid. In alcohol production, a hydrometer is used to test the concentration of sugars in the mash prior to adding yeast and to test the concentration of ethanol in the final distilled product. The hydrometer consists of a sealed glass tube, weighted at one end and with a graduated scale inside. When dropped in a liquid, it floats at a certain height which is read on the scale. This tells you the specific gravity of the liquid which in turn can be used to determine the sugar or alcohol concentration. Some hydrometers have a built-in thermometer to compensate for temperature changes (which affect specific gravity).

malt. Another word for sprouted grain. It usually refers to soaked, sprouted and dried grain, (typically barley) from which the enzymes are used to break starches into sugars in mash preparation.

mash. The name for the soup which is used to make alcohol. The mash consists of whatever raw material is being used (such as corn, wheat, sugar cane, or whey) and water. The raw material is ground, mixed with water, prepared with enzymes where necesary (to break the starches into sugars) and balanced for proper pH (acid level adjusted so the yeast will thrive). The sugar concentration in the mash must not be too high or too low and all conditions must remain sterile to prevent the growth of competing micro organisms—which would reduce the amount of alcohol produced.

methanol. The "other" alcohol commonly used as a motor fuel. Also known as wood alcohol or methyl alcohol. The chemical formula is CH_3OH. It is highly poisonous and is often used as a denaturant for ethanol. It can be made from natural gas, coal wood and accounts for much of the industrial alcohol usage in the United States. Unfortunately, methanol suffers from the same economic and environmental problems as other fuels based on non-renewable resources.

moonshine. Slang for illegally produced drinking liquor. The term has been included here because the techniques for making moonshine can be used to make ethanol fuel.

neat alcohol. Means the same as undilluted or straight alcohol.

pH. A measure of the acidity or alkalinity of a solution. The pH scale ranges from 1 to 14, with 1 being highly acid, 7 neutral and 14 highly alkaline. Pure water has a pH of 7. The pH of the mash is an important aspect of healthy yeast growth. A pH of 4.5 seems to work best. The pH can be measured with special test paper that changes color in response to the pH of the solution or with a pH meter. The pH level can be adjusted by adding acid or alkaline to the mash or by dilluting with water.

phase separation. A problem that sometimes develops when excess water causes gasoline and alcohol to separate. Alcohol absorbs water quite readily but gasoline does not. Therefore, if more than a small percentage of a gasohol mixture contains water, the alcohol, water and gasoline will separate into distinct layers. For this reason, gasohol is blended with 199 proof to 200 proof (anhydrous) ethanol.

185

proof. The alcoholic strength of a liquor or water/alcohol mixture, with 200 proof being pure alcohol and 100 proof being 50 percent water and 50 percent alcohol. If you get confused, try thinking of the proof as twice the percentage of alcohol.

solar still. Distilling apparatus powered by heat from the sun. Solar radiation is used to evaporate the alcohol from the beer. The vapor rises and condenses on a cool surface where it then trickles into a collection trough. Solar stills are typically slow and produce low proof alcohol.

specific gravity. The ratio of the density of a solution with that of pure water. Water has a specific gravity of one. Alcohol, being lighter than water, has a lower specific gravity. Sugar solutions, being heavier than water, have a higher specific gravity. A hydrometer is used to measure the specific gravity, with the scale often calibrated in percentage of sugar or proof of alcohol.

still. Slang for distilling the equipment used to make alcohol. Among moonshiners, the still is the vat which holds the beer. The still is heated so that the alcohol will evaporate from the solution. The vat is sealed except for a small vent, which leads to the condenser where the vapor is turned back into liquid.

stillage. What's left from the mash after the beer has been drained off or strained out. The stillage is a high protein concentrate which can be fed wet to livestock or dried and used as a food supplement (for humans or animals). Dried stillage is the same thing as dried distillers grain.

vacuum still. Distilling apparatus that uses a vacuum to lower the operating temperature of the still for greater energy savings. A vacuum is applied to the vat holding the beer. The reduced pressure lowers the boiling point of the water/alcohol mixture. Less heat is required to boil off the alcohol.

wood alcohol. Another name for methanol or methyl alcohol. The word "wood" refers to the fact that methanol was often made from wood. However, it is now manufactured mostly from natural gas.

yeast. These are the microscopic organisms that make it possible for you to make alcohol. Yeast are fungi that ferment sugars into ethanol alcohol. Once added to the mash, they feed on the sugar to reproduce and grow. As they multiply, they excrete waste products (mainly ethanol and carbon dioxide). This fermentation process continues until the yeast either consume all the sugar or die from the toxic effects of the rising concentration of alcohol and CO_2. Yeast prefer warm temperatures, ranging from 72 degrees to about 90 degrees F. Higher temperatures run the risk of killing the yeast. A pH of about 4.5 is also necessary for rapid growth. Baker's yeast can be used to make alcohol but brewer's strains generally produce higher yields. The final concentration of alcohol in the mash after fermentation has ceased can range from a small percentage to as high as 12 percent or 14 percent, depending on the strain of yeast used, how well it grew and whether or not other organisms invaded the brew.

Appendix
Additional Sources
of Information

Various government agencies, groups and organizations can provide further information about making and using alcohol fuel. The sources provided are by no means an all inclusive list since new groups and organizations seem to spring up overnight in response to the growing interest in alcohol and gasohol. The list can be a starting point if you wish to learn more about making alcohol, alcohol promotion, fuel use or government programs.

In addition to the sources listed, your local public library might have a special alcohol or gasohol information file containing current reports, magazine articles, newspaper stories and research papers on the subject. Some libraries also offer a computerized research service (for a fee) that can give you a summarized printout of current information on any given subject.

Another ready source of information and help is your state department of agriculture. If they do not have any literature for public dissemination or if they are not involved directly in any alcohol programs, they might be able to direct you to the appropriate state agency that can help you.

Another source to explore is the agricultural or agricultural engineering department of state universities and community colleges. You can often locate a "local expert" this way since many universities are actively engaged in alcohol fuel or alternative fuels research. The chemical engineering or chemistry department might also be of assistance in answering questions about still building and distilling columns.

One of the best ways to get a feel for alcohol production is to attend one of the alcohol fuel seminars or workshops that are periodically offered around the country. These learning sessions might be one-day affairs—such as those that have been presented by the Iowa Corn Promotion Board—or they might run several days to a week, such as those sponsored by the Colby Community College (Colby, Kansas) and the National Alcohol Fuel Producers Association.

The best thing about a seminar or workshop is that it puts you in direct contact with people-in-the-know. This gives you an excellent opportunity to ask questions about making and using alcohol. Most of the seminars and workshops will have a working still set up so that you can see for yourself how alcohol is made. The literature which is handed out or sold at these sessions can be extremely helpful, especially when it comes to designing and building your own still. What's more, by rubbing shoulders with others who are interested in alcohol, you can exchange ideas and impressions or perhaps find someone else in your area who is making alcohol fuel.

Some of the workshops will also give you hands-on experience at making alcohol. This can save a lot of lost time and frustration if you're just getting into alcohol production. And it gives you a chance to judge for yourself whether or not alcohol can be a workable solution to your particular energy problem.

Sources

Alcohol—Alternate Fuel Institute
9400 Wisconsin Ave.
Bethesda, MD 20014
(301) 897-5437

Alternate Energy Limited
Route 1
650 Pine St.
Colby, KS 67701
(913) 462-7171

American Soybean Association
P.O. Box 27300
St. Louis, MO 63141
(314) 432-1600

Bureau of Alcohol, Tobacco & Firearms
Department of the Treasury
Washington, D.C. 20226
(202) 566-7268

Distillers Feed Research Council
1435 Enquirer Building
Cincinnati, OH 45202
(513) 621-5985
Domestic Technology Institute
P.O. Box 2043
Evergreen, CO 80439'
(303) 988-3054

Environmental Protection Agency
Fuels Section
EN340
499 South Capitol St. S.W.
Washington, D.C. 20460
(202) 472-9367

Gasohol U.S.A.
10008 East 60th Terrace
Kansas City, MO 64133
(816) 737-0064

Illinois Farm Energy Program
Emmerson Bldg.
Illinois State Fairgrounds
Springfield, IL 62706
(217) 782-6297
International Biomass Institute
1522 K Street N.W.
Suite 600
Washington, D.C. 20005
(202) 223-9136
Iowa Corn Growers Association
200 West Towers
1200-35th St.
West Des Moines, IA 50265
(515) 225-9242

Iowa Corn Promotion Board
402 West Towers
1200 35th Street
West Des Moines, IA 50265
(515) 225-9242

Minnesota Gasohol Commission
P.O. Box 86A
Olivia, MN 56277

National Alcohol Fuel Information Center (Midwest & West)
Southwestern State University
Marshall, MN 56258
(toll free numbers)
1-800-533-5333 (out-of-state callers)
1-800-622-5234 (in-state callers)
National Alcohol Fuel Information Center (South & East)
Nicholls State University
Thibodaux, LA 703011
(toll free numbers)
1-800-535-2840 (out-of-state callers)
1-800-352-2870 (in-state callers)
National Alcohol Fuel Producers Association
P.O. Box 686
Colby, KS 67701
National Association of Wheat Growers
1030 15th St. N.W.
Suite 1030
Washington, D.C. 20005
(202) 466-8630
National Center for Appropriate Technology
P.O. Box 3838
3040 Continental Drive
Butte, MT 59701
(406) 723-5474
National Center for Appropriate Technology (Washington office)
815 - 15th Street
Washington, D.C. 20015
(202) 347-9193

National Gasohol Commission
Executive Bldg.
Suite 5
521 South 14th Street
Lincoln, NE 68508
(402) 475-8055
National Technical Information Service
U.S. Department of Commerce
5285 Port Royal Road
Springfield, VA 22161
(703) 557-4600
Nebraska Agricultural Products
Industrial Utilization Committee
(state agency for alcohol & gasohol)
301 Centennial Mall South
Lincoln, NE 68509'
(402) 471-2941

U.S. Department of Agriculture
14th Street & Independence Avenue S.W.
Washington, D.C. 20250
(202) 447-3631
U.S. Sugar Beet Association
1156 15th Street N.W.
Washington, D.C. 20005
(202) 296-4820

Seminars
Colby Community College
1255 S. Range St.
Colby, KS 67701
(913) 462-3984
Compost Science/Land Utilization
Box 351
Emmaus, PA 18049
(215) 967-4010

Easy Engineering
3353 Larimer Street
Denver, CO 80205
(303) 893-8936

Iowa Western Community College
Council Bluffs, IA
(712) 325-3200

National Alcohol Fuel Producers Association
P.O. Box 686
Colby, KS 67701

Index